Groundwater Vulnerability and Pollution Risk Assessment

T0093478

Selected papers on hydrogeology

24

Series Editor: Dr. Nick S. Robins
Editor-in-Chief IAH Book Series, British Geological Survey, Wallingford, UK

INTERNATIONAL ASSOCIATION OF HYDROGEOLOGISTS

Groundwater Vulnerability and Pollution Risk Assessment

Editors

Andrzej J. Witkowski
Department of Hydrogeology and Engineering Geology,
University of Silesia, Katowice, Poland

Sabina Jakóbczyk-Karpierz
Department of Hydrogeology and Engineering Geology,
University of Silesia, Katowice, Poland

Joanna Czekaj
Department of Modern Technologies and Innovations,
Silesian Waterworks PLC, Katowice, Poland

Dorota Grabala
Department of Hydrogeology and Engineering Geology,
University of Silesia, Katowice, Poland

CRC Press
Taylor & Francis Group
Boca Raton London New York

CRC Press is an imprint of the
Taylor & Francis Group, an **informa** business

A BALKEMA BOOK

Published by:
CRC Press/Balkema
P.O. Box 447, 2300 AK Leiden, The Netherlands
e-mail: Pub.NL@taylorandfrancis.com
www.crcpress.com – www.taylorandfrancis.com

First issued in paperback 2022

© 2020 by Taylor & Francis Group, LLC
CRC Press/Balkema is an imprint of the Taylor & Francis Group, an informa business

No claim to original U.S. Government works

ISBN-13: 978-1-03-240072-3 (pbk)
ISBN-13: 978-0-367-42237-0 (hbk)
ISBN-13: 978-0-367-82292-7 (ebk)

DOI: 10.1201/9780367822927

This book contains information obtained from authentic and highly regarded sources. Reasonable efforts have been made to publish reliable data and information, but the author and publisher cannot assume responsibility for the validity of all materials or the consequences of their use. The authors and publishers have attempted to trace the copyright holders of all material reproduced in this publication and apologize to copyright holders if permission to publish in this form has not been obtained. If any copyright material has not been acknowledged please write and let us know so we may rectify in any future reprint.

Except as permitted under U.S. Copyright Law, no part of this book may be reprinted, reproduced, transmitted, or utilized in any form by any electronic, mechanical, or other means, now known or hereafter invented, including photocopying, microfilming, and recording, or in any information storage or retrieval system, without written permission from the publishers.

For permission to photocopy or use material electronically from this work, please access www.copyright.com (http://www.copyright.com/) or contact the Copyright Clearance Center, Inc. (CCC), 222 Rosewood Drive, Danvers, MA 01923, 978-750-8400. CCC is a not-for-profit organization that provides licenses and registration for a variety of users. For organizations that have been granted a photocopy license by the CCC, a separate system of payment has been arranged.

Trademark Notice: Product or corporate names may be trademarks or registered trademarks, and are used only for identification and explanation without intent to infringe.

Publisher's Note
The publisher has gone to great lengths to ensure the quality of this reprint but
points out that some imperfections in the original copies may be apparent.

Visit the Taylor & Francis Web site at
http://www.taylorandfrancis.com

and the CRC Press Web site at
http//www.crcpress.com

Typeset by Apex CoVantage, LLC

Although all care is taken to ensure integrity and the quality of this publication and the information herein, no responsibility is assumed by the publishers nor the author for any damage to property or persons as a result of operation or use of this publication and/or the information contained herein.

Library of Congress Cataloging-in-Publication Data
Applied for

Contents

Preface

The assessment of the vulnerability of groundwater to contamination has been widely discussed. We continue to observe the intensive development of various methods of vulnerability assessment, both intrinsic and specific (e.g. in terms of pollution by nitrate, pesticides, petroleum substances, sulphates, pharmaceuticals, etc.). Often, issues of specific vulnerability assessment are a fundamental component of every day risk assessment. In this context, the classic term groundwater vulnerability related to groundwater vulnerability to pollution has been significantly extended to cover a number of new aspects related to current geogenic threats stimulated by climate change (e.g. sea water intrusion) as well as new threats resulting from the human activity (problems with so-called emerging contaminants, groundwater intensive drainage, different current and past mining activity, construction activities).

Various modifications of ranking are used successfully in many countries. DRASTIC remains the most popular among the ranking methods for groundwater vulnerability assessment. This method aspires to be the universal method, however, this is not always the case, and it became the basis for the development of new techniques for porous and karst aquifers and for various purposes (regional and local studies) at various scales. Increasingly, the vulnerability assessment takes into account not only simple methods (advection models) determining the travel time of potential contaminants from the ground surface to groundwater, but also more complex models that take into account the role of physical and biological processes in contaminant transport and its attenuation. Unfortunately, there is still controversy about the very concept of vulnerability. Often, authors do not state explicitly whether the vulnerability assessment concerns groundwater resources (aquifer vulnerability) or groundwater sources (for instance well fields). This is important from the point of view of the practical application of vulnerability maps to delineation of aquifer protection zones or wellhead protection zones. These issues have been widely discussed at three international IAH conferences held in Poland in Ustroń spa (in 2004, 2015 and 2018), each devoted to groundwater vulnerability.

An important effect of these conferences was the publication of two volumes of IAH Selected Papers. The first Volume 11 Groundwater Vulnerability Assessment and Mapping was published in 2007. This volume contains papers presented at the conference held in 2004 in Ustroń. Unfortunately, after the conference in 2015, despite many abstracts and numerous interesting presentations sent, the organizers did not receive enough full papers to publish the next volume of IAH Selected Papers. Therefore the papers in the new volume (Vol.24) entitled *Groundwater Vulnerability and Pollution Risk Assessment* were selected from those presented at the last two IAH conferences which was held in 2015 (Groundwater Vulnerability – From Scientific Concept to Practical Application) and in 2018 (New Approaches to

Groundwater Vulnerability). Both conferences were organised by the University of Silesia, IAH, Association of Polish Hydrogeologists and UNESCO IHP.

This volume contains 15 chapters presented at both conferences and has been divided into four main parts: New approaches to groundwater vulnerability (Chapter 1–4), Factors affecting vulnerability assessment – from scientific concept to practical application (Chapter 5–8), Comparison and validation of different methods of groundwater vulnerability assessment for different groundwater systems (Chapter 9–11), Groundwater vulnerability mapping – examples of different national approaches (Chapter 12–15).

The Editors would like to express sincere thanks to Nick Robins (Editor-in-Chief IAH Book Series) and Janjaap Blom (Senior Publisher at Taylor & Francis) for excellent cooperation, understanding and kind help in publishing both volumes.

The Editors would like to express particularly warm acknowledgments to the authors for their contributions as well as for their patience and understanding with regard to publication. We also would like to thank scientific reviewers for their careful reviews and for their efforts in the linguistic correction of some manuscripts.

The Editors would like to address special thanks to UNESCO International Hydrological Program and personally to Dr. Alice Aureli for financial support. This support made it possible for scientists from developing countries to participate in the conference and present their papers. Without this help, a number of papers contained in this volume could not be published.

Andrzej J. Witkowski

Contributors

Abdulhamid A. – Faculty of Earth and Environmental Sciences, Bayero University, Kano, Nigeria

Ahmed M.S. – Kaduna Refining and Petrochemical Company Limited (KRPC/NNPC), Kaduna, Nigeria

Atanacković N. – University of Belgrade, Faculty of Mining and Geology, Department of Hydrogeology, Belgrade, Serbia

Aureli A. – UNESCO – Chief of Section Groundwater Systems and Water for Human settlements, Division of Water Sciences – International Hydrological Programme (IHP), 75732 Paris, France

Badamasi M.M. – Faculty of Earth and Environmental Sciences, Bayero University, Kano, Nigeria

Boyraz U. – Department of Civil Engineering, Istanbul University-Cerrahpaşa, Istanbul, Turkey

Carrubba S. – Geoprospezioni.it, 90018 Termini Imerese (PA), Italy

Civita M.V. – Professor of Applied Hydrogeology (ret) – Politecnico di Torino. Head of Italian Program on Groundwater Contamination GNDCI-CNR

David K. – School of Minerals and Energy Resource Engineering and Connected Water Initiative Research Centre, UNSW Australia, Sydney, NSW, Australia

De Filippi F.M. – Department of Civil, Building and Environmental Engineering Sapienza, University of Rome, via Eudossiana, 18–00184 – Rome, Italy

Dragišić V. – University of Belgrade, Faculty of Mining and Geology, Department of Hydrogeology, Belgrade, Serbia

Dragon K. – Adam Mickiewicz University in Poznan, Institute of Geology, Department of Hydrogeology and Water Protection, Poznań, Poland

Ferranti F. – Department of Civil, Building and Environmental Engineering Sapienza, University of Rome, via Eudossiana, 18–00184 – Rome, Italy

Fister V. – EPTB Saône et Doubs, 36 rue Saint-Laurent, 25290 Ornans, France; LOTERR, UFR SHS, Ile du Saulcy, CS 60228, 57045 Metz cedex 1, France

François D. – LOTERR, UFR SHS, Ile du Saulcy, CS 60228, 57045 Metz cedex 1, France

Gille E. – LOTERR, UFR SHS, Ile du Saulcy, CS 60228, 57045 Metz cedex 1 – France

Herbich P. – Polish Geological Institute – National Research Institute, Warsaw, Poland

Hickey C. – Geological Survey of Ireland, Ireland

Jemcov I. – University of Belgrade, Faculty of Mining and Geology, Department of Hydrogeology, Belgrade, Serbia

Jóźwiak K. – Polish Geological Institute – National Research Institute, Warsaw, Poland

Kazezyılmaz-Alhan C.M. – Department of Civil Engineering, Istanbul University-Cerrahpaşa, Istanbul, Turkey

Kelly C. – Tobin Consulting Engineers, Ireland

Lagod M. – "Formerly of UNESCO – Division of Water Sciences – International Hydrological Programme (IHP)" UN Environment Programme/Mediterranean Action Plan (UNEP/MAP), 18019 Athens, Greece

Lee M. – Geological Survey of Ireland, Ireland

Lemieux J.-M. – Department of Geology and Geological Engineering; Université Laval, Québec, QC, Canada G1V 0A6

Losson B. – LOTERR, UFR SHS, Ile du Saulcy, CS 60228, 57045 Metz cedex 1 – France

Marchetto M. – Agence de l'eau Rhin-Meuse (AERM), Route de Lessy, 57160 Rozérieulles, France

Meehan R. – Talamhireland Consultancy, Ireland

Mikołajków J. – Polish Geological Institute – National Research Institute, Warsaw, Poland

Mitra R. – School of Mining Engineering, University of the Witwatersrand, Johannesburg, South Africa

Molson J. – Department of Geology and Geological Engineering; Université Laval, Québec, QC, Canada G1V 0A6

Nowamooz A. – BluMetric Environmental, Montreal, QC, Canada H2Y 1N3

Nidental M. – Polish Geological Institute – National Research Institute, Warsaw, Poland

Oke S.A. – Unit for Sustainable Water and Environment, Department of Civil Engineering, Central University of Technology Bloemfontein, South Africa

Okońska M. – Institute of Physical Geography and Environmental Planning, Adam Mickiewicz University in Poznań, Krygowskiego 10, 61–680 Poznań, Poland

Pietrewicz K. – Institute of Physical Geography and Environmental Planning, Adam Mickiewicz University in Poznań, Krygowskiego 10, 61–680 Poznań, Poland

Roy N. – Rapille-dessus 13, 1312 Eclépens, Switzerland

Sappa G. – Department of Civil, Building and Environmental Engineering Sapienza, University of Rome, via Eudossiana, 18–00184 – Rome, Italy

Tanko A.I. – Faculty of Earth and Environmental Sciences, Bayero University, Kano, Nigeria

Timms W. – Faculty of Science Engineering & Built Environment, Deakin University, Melbourne, Victoria, Australia

Van Stempvoort D. – Environment and Climate Change Canada, National Water Research Institute, Burlington, ON, Canada L7R 4A6

Vermeulen D. – Institute for Groundwater Studies, University of Free State Bloemfontein, South Africa

Williams N.H. – Geological Survey of Ireland, Ireland

Woźnicka M. – Polish Geological Institute – National Research Institute, Warsaw, Poland

Živanović V. – University of Belgrade, Faculty of Mining and Geology, Department of Hydrogeology, Belgrade, Serbia

About the editors

Andrzej J. Witkowski, PhD, DSc, is an associate professor at the University of Silesia in Poland and Head of the Department of Hydrogeology and Engineering Geology. His academic work has included studies on hydrogeology, groundwater protection, vulnerability and monitoring. He is the author or coauthor of 162 publications in the area of hydrogeology and groundwater protection and author or coauthor of about 130 unpublished projects and reports. He was co-editor of IAH Selected Papers on Hydrogeology, Vol.11. – "Groundwater vulnerability assessment and mapping". He is editor in Chief of the Polish journal "Hydrogeologia" and member of the Editorial Board of the Slovak journal "Podzemna Voda". Professional affiliations: IAH – International Association of Hydrogeologists (President of the Polish National Chapter, 2010–2016), IMWA – International Mine Water Association (Vice President, 2000–2003; President, 2003–2008; Honorary President – since 2008), SHP- Association of Polish Hydrogeologists (President – since 2011), SAH- Slovenska Asociácie Hydrogeológov.

Sabina Jakóbczyk-Karpierz, PhD, is a hydrogeologist. She has worked at the Department of Hydrogeology and Engineering Geology, Faculty of Earth Sciences, University of Silesia in Katowice (Poland) since 2008. Her research interests are groundwater geochemistry, geochemical modelling and application of environmental tracers in groundwater dating and protection. She has been involved in several European projects concerning groundwater quality and management of groundwater resources. She was reviewer for such Journals as: "Journal of Hydrology", "International Journal of Environmental Research and Public Health", "Water". She is a member of the IAH and SHP – Association of Polish Hydrogeologists.

Joanna Czekaj, MSc, has a 8-year experience in hydrogeology. Her main field of research is groundwater-surface water interaction, especially the problem of groundwater interaction with artificial, drinking water reservoirs. Her interest is focus on GIS, water resources modelling and its application in sustainable water resources management. One of the most important aspects of Joanna's work is research implementation under R&D activities. Currently, Joanna is working at R&D department in one of the biggest water supplying company in Poland – Silesian Waterworks PLC and her responsibilities include managing of PROLINE-CE and boDEREC-CE projects, co-funded by Interreg CENTRAL EUROPE program.

Dorota Grabala, MSc, is researcher at the Department of Hydrogeology and Engineering Geology, University of Silesia (Poland). She is involved in regional hydrogeological studies and problems of groundwater protection and monitoring. She has been involved in several Polish and international projects on groundwater pollution and management (WOKAM – World Karst Aquifer Mapping, INTERREG CE).

New approaches to groundwater vulnerability

ACVM (Aquifer Comprehensive Vulnerability Mapping) – a new method for evaluating coastal aquifer vulnerability based on a wide concept of aquifer vulnerability

S. Carrubba, A. Aureli & M. Lagod

I Introduction

ACVM is a new method that was conceived, developed and applied in the area of Ghar El Melh in Tunisia in the context of the Global Environment Facility (GEF)/UN Environment Programme (UNEP)/Mediterranean Action Plan (MAP) Strategic Partnership for the Mediterranean Sea Large Marine Ecosystem (MedPartnership, 2009–2015). UNESCO's International Hydrological Programme (IHP) was responsible for the execution of MedPartnership Subcomponent 1.1 on "Management of Coastal Aquifers and Groundwater", which provided the framework for this activity. ACVM method was developed in a sedimentary coastal aquifer but it can be applied in every kind of coastal aquifer.

2 The application of the vulnerability concept to coastal aquifer

The evolution of the aquifer vulnerability concept started in the late 1960s, when hydrogeologists began formulating definitions to describe the notion. The following definitions of vulnerability illustrate how the concept of vulnerability has evolved since then:

*   Vulnerability is an intrinsic property of a groundwater system that depends on the sensitivity of that system to human and/or natural impacts;

    (Vrba and Zaporozec, 1994)

*   Vulnerability is the intrinsic characteristics which determine the sensitivity of various parts of an aquifer to being adversely affected by an imposed contaminant load;

    (Foster, 1987)

*   Vulnerability on the human time scale is an unchanging natural intrinsic property of the unsaturated and saturated parts of a groundwater system and depends on the ability or inability of this system to cope with natural processes and human impacts.

    (Vrba, 1991)

Therefore, the concept of groundwater vulnerability is based on the assumption that the physical environment provides some degree of protection to groundwater against the natural and human impacts, especially with regard to contaminants entering the subsurface environment. It follows that some land areas are more vulnerable to groundwater contamination than others. Accordingly, the vulnerability map can be used to indicate the different levels of protection that the natural environment provides to groundwater resources.

The definitions of vulnerability provided above essentially describe vertical vulnerability, in other words, the vulnerability against a pollutant penetrating the aquifer through the ground surface.

However, in coastal aquifers, groundwater can be threatened by pollutants originating from the ground surface as well as by salt water entering from the sea or coastal lagoons that can penetrate the aquifer through hydrogeological connections. Therefore, in coastal aquifers it is necessary to consider a new, wider concept of vulnerability, a sort of "comprehensive vulnerability" (Figure 1.1) that simultaneously considers these two aspects of aquifer vulnerability. These will be referred to as:

- "vertical vulnerability" for the vulnerability of contamination threats from land-based pollution
- "horizontal vulnerability" for the vulnerability related to the salt water intrusion phenomena.

In coastal areas the two phenomena are linked and interact with one another. In fact, salt water can heavily influence the chemical reaction between pollutants and rocks, and among pollutants themselves.

The objective of this study was to develop an experimental methodology for the evaluation of the comprehensive vulnerability of aquifers in coastal areas. The result of this study is the ACVM methodology, named for the acronym of Aquifer Comprehensive Vulnerability Mapping (Carrubba, 2014a, 2014b; Carrubba *et al.*, 2015a).

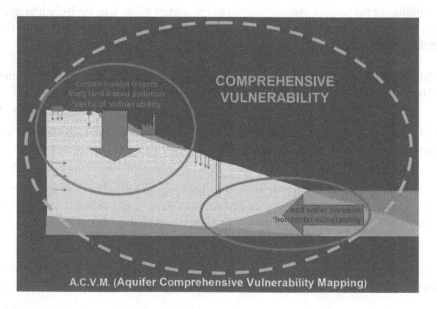

Figure 1.1 Vulnerability mapping in Coastal area.

3 Coastal aquifer vulnerability mapping using the ACVM (Aquifer Comprehensive Vulnerability Mapping) methodology

ACVM is a new methodology that can simultaneously describe many vulnerability aspects of a territory, including vertical and horizontal vulnerability. It can also be extended to other aspects of vulnerability, including sea level rise.

Rising sea levels – one of the effects of climate change – can decrease groundwater quality in the long-term, increasing both vertical and horizontal vulnerability. On the one hand, a modest increase in sea levels can result in a reduction of the thickness of the unsaturated zone, increasing the aquifer's vertical vulnerability. On the other hand, rising sea levels subject coastal aquifers to increased horizontal vulnerability and can result in significant sea water intrusion.

This additional vulnerability aspect can be incorporated in the ACVM method as the potential impact on groundwater quality of variation in sea levels.

Therefore, the ACVM method can also be used to establish a map that indicates the parts of an aquifer that are more or less vulnerable to global climate change (Carrubba *et al.*, 2015b).

The ACVM method establishes a new conceptual approach to evaluating aquifer vulnerability, is easy to use and can be applied with low-cost data. Of course, the accuracy of the results will depend on the quality of the data available.

4 Introduction to the concept of comprehensive vulnerability

The ACVM (Aquifer Comprehensive Vulnerability Mapping) methodology is based on a new, expanded concept of vulnerability mapping that is derived from the concept of intrinsic aquifer vulnerability itself. Vulnerability is an intrinsic characteristic of an aquifer because it is linked to the intrinsic characteristics of the aquifer that provide a defence against an external threat, such as a pollutant, a contaminant or an external event that can impact water quality. Therefore, when developing an aquifer vulnerability map, it is important to consider the intrinsic characteristics of an aquifer that offer protection against the potential external threats in the short- and long-term (Carrubba *et al.*, 2016).

The schematic process for using the ACVM (Aquifer Comprehensive Vulnerability Mapping) is constituted by the following steps:

- Step 1: Mapping **vertical vulnerability** using the most appropriate method depending on data availability and their distribution using a numerical scale.
- Step 2: Mapping **horizontal vulnerability** associated with salt water intrusion using the same numerical scale.
- Step 3: Evaluating the potential impact on groundwater quality due to the variation of sea levels, **sea level rise vulnerability**.
- Step 4: Merging Vertical, Horizontal and Sea level rise vulnerability maps in order to map **comprehensive vulnerability in the short- and long-term**.

The ACVM (Aquifer Comprehensive Vulnerability Mapping) method requires a numerical vulnerability scale where 0 is the lowest vulnerability class and the highest number of the scale corresponds to the severest vulnerability class.

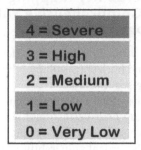

Figure 1.2 Meaning of the Numeric Vulnerability Scale

It is not important how many classes there are but the numerical scales must be the same for all aspects of aquifer vulnerability. In this study a numerical vulnerability scale constituted by 5 classes is used (Figure 1.2). In particular 0 indicates very low vulnerability, 1 indicates low vulnerability, 2 indicates a medium vulnerability area, 3 indicates high vulnerability and 4 indicates areas where vulnerability is considered severe.

4.1   Step 1: mapping vertical vulnerability

The vertical intrinsic vulnerability of an aquifer is linked to its natural characteristics that contribute to its defence from pollution, such as the depth to the water table, the infiltration rate, the attenuation action of the unsaturated zone, the soil attenuation capacity, the geological formation and the aquifer types, the hydraulic conductivity, the slope of the surface and other natural characteristics that can influence the aquifer natural protection.

ACVM is compatible with all standard methods for evaluating vertical vulnerability, and can apply the most appropriate method for evaluating vertical vulnerability depending on the amount of available data and their quality and distribution.

If a large amount of data are available and well distributed over the study area, the study area can be divided into a grid and a parametric method can be used to calculate and map the vulnerability. Conversely, if few data are available or if the data are not well distributed over the study area, it is preferable to divide the territory into homogeneous areas that have the same vulnerability grade based on a few main characteristics of the hydrogeologic complex.

4.2   Step 2: mapping horizontal vulnerability

While there are several standardised methodologies for evaluating vertical vulnerability, the same is not true for the evaluation of horizontal vulnerability. In fact, the studies consulted on salt water intrusion only demonstrate methodologies for modelling the state of salt water intrusion. In this way, they study the damage and not aquifer vulnerability.

The first objective of this study was to develop and to propose a methodology for studying the vulnerability of an aquifer to salt water intrusion, starting from the classic definition of vulnerability. When evaluating and mapping aquifer vulnerability to the horizontal threat of salt water intrusion, it is important to consider the intrinsic characteristics of the aquifer that protect it from this threat.

The main defence against salt water intrusion is specific energy, which is linked to the head of groundwater in the aquifer. For this reason, an aquifer with a level of fresh groundwater

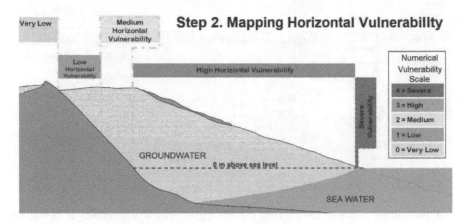

Figure 1.3 Schematic of the designation of vulnerability classes based on the level of fresh groundwater, for the horizontal vulnerability mapping methodology

that is several meters above sea level has a greater capacity to resist the influence of sea water intrusion than an aquifer with a level of fresh groundwater that is at or near sea level.

Figure 1.3 provides a schematic of a cross-section of the freshwater/salt water interface in a typical coastal zone, and shows the vulnerability classes for the evaluation of horizontal vulnerability to the threat of sea water intrusion.

Areas of an aquifer where the level of fresh groundwater is 0 m above sea level (asl) or lower can be assigned a vulnerability grade of **severe**, since in these places the aquifer has a relatively low specific energy, and in turn a weakened ability to resist the threat of salt water intrusion.

Areas of an aquifer where the level of fresh groundwater is greater than 0 m asl and where the bottom of the aquifer has an elevation less than or equal to 0 m asl can be assigned a vulnerability class of high.

Areas where the base of an aquifer is above sea level can be assigned a vulnerability class of medium, since there is still a possibility that human activities can bring about salt water intrusion in these locations, for example as a result of the application of moderately saline irrigation water to the land.

Areas of the aquifer further inland that can represent feeding areas for the aquifer can be assigned a vulnerability class of low, while in those parts of the territory where aquifer has no connection with salt water, the vulnerability class will be very low.

On this basis, five vulnerability classes have been established for the designation of aquifer vulnerability to salt water intrusion. It is important to underline that the limits of the zones of different horizontal vulnerability can be established with inexpensive data by modelling a large number of geological sections based on a good geological surveying and by performing hydrogeological correlations using well level data.

4.3 Step 3: mapping sea level rise vulnerability

Many studies agree that sea water levels will rise in the coming decades. Obviously, this will affect coastal areas, and especially flat coastal areas where shorelines will recede inland relatively quickly.

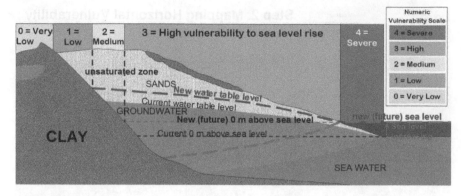

Figure 1.4 Method for evaluating the potential impact on groundwater vulnerability of variation in sea levels using ACVM

While the effects of sea level variations on land surfaces are easy to forecast, the same is not true for their effects on aquifer vulnerability. Rising sea levels will push the groundwater layer upwards towards the land surface, consequently reducing the thickness of the unsaturated zone and increasing the aquifer's vertical vulnerability. On the other hand, sea level rise will increase the thickness of the salt water layer, which will in turn increase the specific energy of the body of salt water, resulting in increased vulnerability to seawater intrusion. In view of this, there is a need for a methodology that can provide guidance for predicting the variation in aquifer vulnerability associated with changing sea levels.

The ACVM method can incorporate many aspects of aquifer vulnerability, so it can be used to study the potential impact on groundwater quality of the variation of sea levels and the resulting increase in comprehensive vulnerability. This methodology can be used with simple or sophisticated numerical models depending on data availability.

The approach for delineating the different classes of horizontal vulnerability is based on the reference point of 0 m above sea level. If the sea level rises, a map can be established based on the new reference point for the sea level, indicating the areas of the aquifer that are likely to experience an increased level of comprehensive vulnerability.

Like the external threats of pollution from land-based activities and seawater intrusion, sea level rise can also be considered as an external threat that can have a negative potential impact on groundwater quality. Figure 1.4 sets out an approach for evaluating aquifer vulnerability to sea level rise using five vulnerability classes, the same number of vulnerability classes as are used in the examples for horizontal and vertical vulnerability.

- **Severe vulnerability to sea level rise:** the areas of the coast that are expected to be submerged as a result of sea level rise will have a severe vulnerability. It is important to underline that the boundaries of this area will be defined not only topographically, but also in terms of hydrogeological and geomorphological aspects. This concept is more evident in flat coastal areas where sea level rise can produce the inundation of depressed areas using river beds and channels or simply by passing coastal barriers or dunes system. In this case, the dunes system will also have severe vulnerability to sea level rise although their topographic elevation could be higher than the new sea level.
- **High vulnerability to sea level rise:** the parts of the aquifer where the bottom of the aquifer will be under the new (future) sea level and the water table level will increase

(because of the sea level rise) will have a high vulnerability to sea level rise. In these areas the thickness of the unsaturated zone will be reduced because of the increase of water table level, and this will increase vertical vulnerability since a pollutant coming from the surface will easily reach the groundwater. Salt water will also be naturally present on a wider area of the aquifer, increasing horizontal vulnerability.

- **Medium vulnerability to sea level rise:** the areas where the current elevation of the bottom of the aquifer will still be above the new (future) sea level will have a medium vulnerability to sea level rise. The main impact will be on vertical vulnerability because of the reduction of the thickness of the unsaturated zone.
- **Low vulnerability to sea level rise:** in all parts of the aquifer further inland that can represent feeding areas for the aquifer, the vulnerability to sea level rise will be low.
- **Very low vulnerability to sea level rise:** in all parts of the territory where aquifers are not connected with the sea also with the new (future) sea level, the vulnerability to sea level rise will be very low.

Using this approach, a map of aquifer vulnerability to sea level rise can be established using the numerical values associated with the five vulnerability classes. This map can be used in two ways: as a stand-alone map of aquifer vulnerability to sea level rise, or alternatively as a basis for incorporating the dimension of vulnerability to sea level rise into the assessment of comprehensive vulnerability.

4.4   Step 4: merging vertical, horizontal and sea level rise vulnerability maps in order to map comprehensive vulnerability in the short- and long-term

The vertical and horizontal sea level rise vulnerability were mapped using the same numeric vulnerability scale (ranging from 0 to 4). Each map is composed of polygons that represent areas with the same class of the vulnerability aspect of the aquifer.

The fourth step consists of overlaying the maps and in intersecting all the polygons of each vulnerability map to obtain a new map where each polygon has a new numeric vulnerability class that is obtained by adding the numeric vulnerability class assigned to each intersected polygon. The resulting numeric value gives the comprehensive vulnerability of that intersected polygon.

Conceptually, this is akin to overlaying maps to illustrate the cumulative effect of many types of vulnerability in a given area. It is still necessary for the maps that describe every type of vulnerability to have the same number of vulnerability classes.

The innovative aspect of this methodology is its ability to provide a visual representation of aquifer vulnerability to several types of external threats, by establishing a cumulative parameter of comprehensive vulnerability and presenting this on a map.

Depending on which aspect of aquifer vulnerability that is being considered, the resulting map will be a short-term or long-term comprehensive vulnerability map.

If only vertical and horizontal vulnerability are merged, the resulting map will be a short-term comprehensive vulnerability map. Alternatively, if vertical, horizontal and sea level rise vulnerability maps are merged the map will be a long-term comprehensive vulnerability map.

Since the parameter for comprehensive vulnerability is calculated by taking the sum of the values for vertical, horizontal and sea level rise vulnerability classes (ranging from 0 to 4) for a given area, the resulting value for the short-term comprehensive vulnerability will be associated with one of nine possible vulnerability classes, with numeric values ranging from 0 to 8, where 0 indicates very low comprehensive vulnerability and 8 indicates

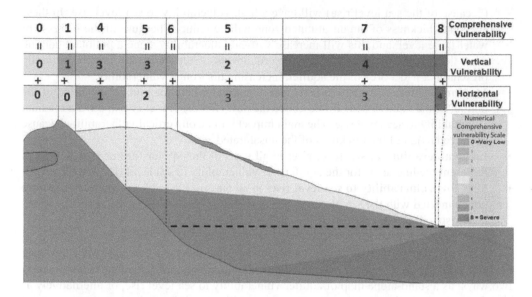

Figure 1.5 Schematic process for the calculation of short-term comprehensive vulnerability. Vertical and horizontal vulnerability maps are overlaid and all the polygons that delimit vulnerability classes of each map are intersected. Each intersected polygon has a comprehensive vulnerability parameter that is the sum of vertical and horizontal vulnerability value.

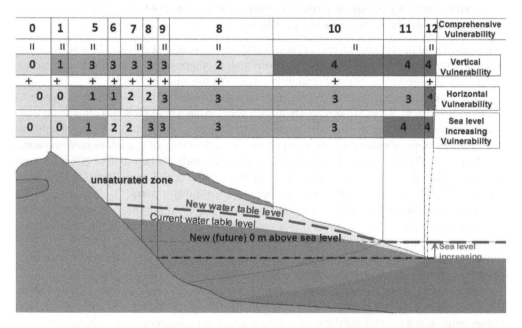

Figure 1.6 Schematic process for the calculation of long-term comprehensive vulnerability. Vertical, horizontal and sea level rise vulnerability maps are overlaid and all the polygons that delimit vulnerability classes of each map are intersected. Each intersected polygon has a comprehensive vulnerability parameter that is the sum of vertical, horizontal and sea level rise vulnerability value.

severe comprehensive vulnerability. While, the resulting value for the long-term comprehensive vulnerability will be associated with one of thirteen possible vulnerability classes, with numeric values ranging from 0 to 12, where 0 indicates very low comprehensive vulnerability and 12 indicates severe comprehensive vulnerability.

The schematic diagrams in Figure 1.5 and 1.6 demonstrates the approach for calculating and mapping the new comprehensive vulnerability parameter. The numbers in these diagrams represent the numeric vulnerability class from the vertical, horizontal and sea level rise vulnerability maps. The comprehensive vulnerability for an area is obtained by intersecting all the polygons that constitute the various aspects of vulnerability (Figure 1.7).

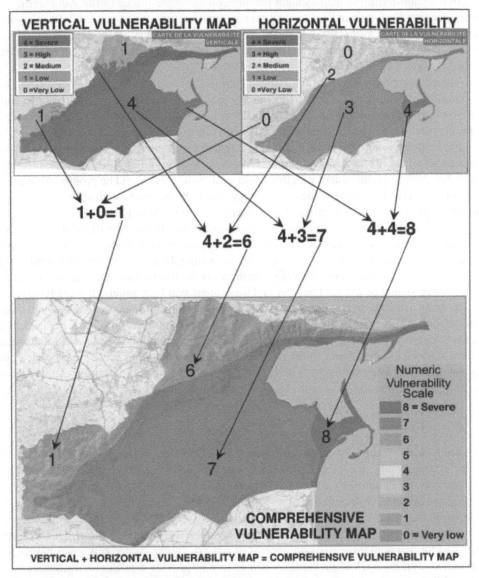

Figure 1.7 Schematic of the approach for the calculation of comprehensive vulnerability. In this figure the short-term comprehensive vulnerability map is calculated intersecting the polygons of vertical and horizontal vulnerability maps. Each polygon of the comprehensive vulnerability map has a comprehensive vulnerability parameter that is the sum of the vertical and the horizontal vulnerability value of each intersected polygon.

5 Management recommendations

Comprehensive vulnerability mapping is a new tool that uses a single value to simultaneously describe several aspects of coastal aquifer vulnerability. Furthermore, it can be an integrated vulnerability map if the locations of potential pollution sources and the target are for possible groundwater pollution are plotted on the comprehensive vulnerability map. Consequently, it can be a valuable resource for land use management, as it resolves the vulnerability parameters into management recommendations for the short- and long-term (Carrubba, 2017a, 2017b).

In the short-term, the management recommendations are linked to the vulnerability of an aquifer to pollution from land-based activities (vertical vulnerability) and from salt water intrusion (horizontal vulnerability). But as a basis for sound planning for the exploitation of an area, the long-term management recommendations also consider the impact of sea level rise as a result of climate change (Carrubba et al., 2017).

The comprehensive vulnerability maps, both in the short- and long-term, are based on the information contained in each vulnerability map. Consequently, it is important to consider each of the two or three derived maps when viewing the comprehensive vulnerability map, in order to understand which component of vulnerability is prevalent.

For instance, areas where the aquifer has a severe comprehensive vulnerability class are those areas where all the vulnerability aspects have severe vulnerability classes, meaning that in that areas all human activities that could emit pollutants should be prohibited. If these activities are already present in these areas, they should be equipped with appropriate pollution prevention systems. Similarly, groundwater abstraction should be controlled by regulation to prevent significant modifications of groundwater levels, which will reduce the risk of salt water intrusion in the aquifer. Furthermore, this part of the territory will be submerged if sea levels rise, so it is necessary to consider the vulnerability of the land to flooding and submersion over the long-term. Therefore, in these areas all activities that have an impact on groundwater should be avoided entirely and these areas will be submerged if sea level rises.

In areas with other comprehensive vulnerability classes, management recommendations should consider the vulnerability aspects that are prevalent and if the aquifer has some means of defence for that threat. For example, it is possible that in a given area groundwater is fully exposed to a pollutant coming from the land surface but it has some means of defence against salt water intrusion and sea level rise. Consequently, the exploitation of this area is possible in the short- and long-term, but will need careful management.

6 Conclusion

This study developed the ACVM methodology for mapping various aspects of aquifer vulnerability using only one output parameter, which is called comprehensive vulnerability. The components of aquifer vulnerability that were considered in this study are vertical vulnerability, horizontal vulnerability of an aquifer and salt water intrusion and sea-level rise vulnerability in the context of global climate change. Horizontal vulnerability and sea-level rise vulnerability were not adequately addressed by previous methodologies, so in this study a method for mapping these two components of coastal aquifer vulnerability was developed and included in the ACVM method. ACVM is able to provide a rating for comprehensive vulnerability that simultaneously considers vertical vulnerability, horizontal vulnerability and sea level rise vulnerability.

This methodology is particularly relevant to coastal areas, and the result is a short-term comprehensive vulnerability map and a long-term comprehensive vulnerability map. Moreover, the management recommendations are attached to the map and provide guidance for decision makers on measures that can be taken to protect these vulnerable groundwater resources in both the short- and long-term, by taking protective measures to reduce the pollution threat of existing activities, or by placing restrictions on the kinds of new activity that are allowed in certain areas.

Therefore, the ACVM method can be a valuable tool for sound land use planning and for safeguarding the environment and groundwater resources against pollution and overexploitation in the short- and long-term.

Finally, the ACVM methodology constitute a new approach for assessing coastal aquifer vulnerability, using a new wider concept of vulnerability. It is being applied to various Mediterranean test sites to demonstrate its potential for widespread use.

References

Carrubba, S. (2014a) *Demonstration of Vulnerability Mapping Techniques in the Ghar El Melh Coastal Aquifer*. MedPartnership Steering Committee Report, Hammamet, Tunisia, 17–20 February.

Carrubba, S. (2014b) *UNESCO-IHP Vulnerability Mapping Demonstration Project at the Ghar El Melh Coastal Aquifer in Tunisia*. Third UNESCO/GEF IW:LEARN Groundwater Integration Dialogue Managing Groundwater in Coastal Areas and SIDS. Report, Athens, Greece, 6–7 May.

Carrubba, S. (2017a) *Vulnerability Mapping of the Ghar El Melh Coastal Aquifer in Tunisia*, published by UNEP-MAP, UNESCO-IHP (2017). Main Hydro(geo)logical Characteristics, Ecosystem Services and Drivers of Change of 26 Representative Mediterranean Groundwater-Related Coastal Wetlands. Strategic Partnership for the Mediterranean Sea Large Marine Ecosystem (MedPartnership), Paris. SC-2015/WS/xx. Available from: http://unesdoc.unesco.org/images/0024/002470/247097e.pdf.

Carrubba, S. (2017b) *Cartographie Des Vulnérabilités De L'aquifère Côtier De Ghar El Melh En Tunisie*, published by PNUE-PAM, UNESCO-PHI (2017). Cartographie des vulnérabilités de l'aquifère côtier de Ghar El Melh en Tunisie. Partenariat stratégique pour le grand écosystème marin de la mer Méditerranée (MedPartnership). Paris. SC-2017/WS/xx. Available from: http://unesdoc.unesco.org/images/0024/002470/247097f.pdf.

Carrubba, S., Lagod, M. & Stephan, R. (2015a) A.C.V.M. (Aquifer Comprehensive Vulnerability Mapping) a new method for evaluating coastal aquifer vulnerability. *Proceedings of the International Conference on Groundwater Vulnerability From Scientific Concept to Practical Application*, Ustron, Poland, 25–29 May.

Carrubba, S., Lagod, M. & Stephan, R. (2015b) A.C.V.M.: A method to evaluate climate change impacts on coastal aquifers. *Proceedings of the Twelfth International Conference on Mediterranean Coastal Environment, Varna, Bulgaria. Erdal Ozhan Editor, 6–10 October*, Volume 2, pp. 515–526.

Carrubba, S., Lagod, M. & Stephan, R. (2016) Short and long term coastal aquifer comprehensive vulnerability mapping: The ACVM method a valuable tool for groundwater managing plan. *[Lecture] 43rd IAH Congress*, Montpellier, France, 25–29 September.

Carrubba, S., Aureli, A. & Lagod, M. (2017) ACVM (Aquifer Comprehensive Vulnerability Mapping): A new tool for assessing potential impacts on the sustainability of coastal aquifers and habitats. *[Lecture] 44th IAH Congress*, Dubrovnik, Croatia, 25–29 September.

Foster, S.S.D. (1987) Fundamental concepts in aquifer vulnerability, pollution risk and protection strategy. In: van Duijvenbooden, W. & van Waegeningh, H.G. (eds.) *Vulnerability of Soil and Groundwater to Pollutants*. TNO Committee on Hydrological Research, Proceedings and Information No. 38, The Hague. pp. 69–86.

Vrba, J. (1991) *Mapping of Groundwater Vulnerability*. International Association of Hydrogeologists, Ground Water Protection Commission, Unpublished Working Paper for Meeting in Tampa, FL, USA.

Vrba, J. & Zaporozec, A. (1994) *Guidebook on Mapping Groundwater Vulnerability, International Contributions to Hydrogeology*, Volume 16. UNESCO/International Association of Hydrogeologists, Hanover, Germany.

Chapter 2

Assessment of the intrinsic vulnerability of the Rhine-Meuse basin limestone aquifers

V. Fister, B. Losson, D. François, E. Gille &
M. Marchetto

Introduction

Limestone aquifers are highly vulnerable to contamination due to high flow velocities (Mangin, 1975) and weak filtration capacities. To express the karstic vulnerability of limestone aquifers, numerous methods of multi-criteria mapping have been developed, such as the EPIK method initiated in the Swiss Jura (Dörfliger and Zwahlen, 1998). By overlaying several layers of geographic information on a predefined grid, an overall vulnerability index can be calculated on each cell. Other multi-criteria mapping models were then developed: the RISKE (Pételet-Giraud et al., 2000), the RISKE 2 (Pranville et al., 2008), the KARSTIC (Davies et al., 2002) the PI (Goldscheider, 2005) or more recently the PaPRIKa method (Kavouri et al., 2011). Inspired by this type of multi-criteria mapping, the Vac arm method we are now going to developed assesses the vulnerability of the main limestone aquifers of the Rhine-Meuse basin. The Rhine-Meuse Basin is one of the six major river basins in France and covers an area of 32700 km² in the Northeast of France (Figure 2.1). It includes ancient massifs (the Vosges and the Ardennes) and sectors of plains and plateaus dependent on the Paris basin and the Upper Rhine Graben. In the Rhine-Meuse basin, the drinking water supply is primarily ensured by groundwater bodies. The main ones are the alluvial aquifers of the major rivers (the Rhine, the Moselle, the Meuse), the Vosges sandstone aquifer and the limestone aquifers. Three limestone aquifers are mainly exploited, respectively contained in the Ladinian, Bathonian- Bajocian and Oxfordian plateaus.

The exploitation of these three aquifers represents an annual volume of 60 million m3, that is approximately 10 % of the underground volume taken on the scale of the Rhine-Meuse basin. In addition, the limestone aquifers of the Rhine-Meuse basin were assessed as highly affected by nitrate and phytosanitary pollution in the inventory drawn up, 2004, for the European Water Framework Directive (AERM, 2009). Consequently, these different water masses were defined as "at risk" and their return to a good ecological condition has been postponed to the year 2027 (2015 for other groundwater bodies). One of the factors governing the pollution risk is the development of karst phenomena. The recent establishment of a database (BDIKARE) identifying the karst phenomena of the Rhine-Meuse basin currently allows the implementation of an identification tool of vulnerability in calcareous environment. By superimposing both functioning criteria (infiltration, endokarstification) and structural criteria (the lithology and thickness of the unsaturated zone and the lithology of the saturated zone), vulnerability maps are obtained. The purpose of these maps is to provide a tool, intended to water's stakeholders, which contributes to a better knowledge of the functioning and to a better conservation of resources.

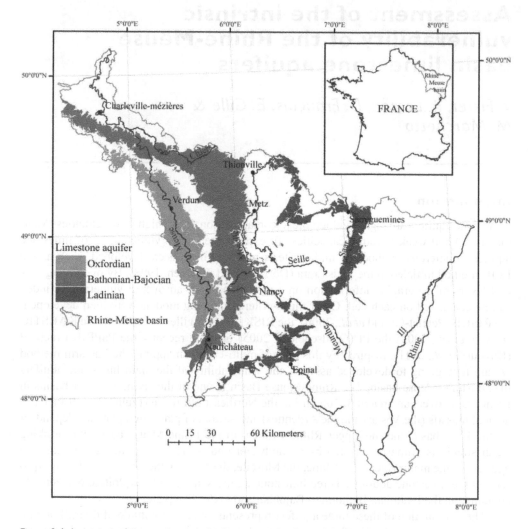

Figure 2.1 Location of the study area: the limestone aquifers of the Rhine-Meuse basin

The characteristics of the limestone aquifers of the Rhine-Meuse basin

The criteria used in the VACARM method are set by the characteristics that should be present. Unlike many high hydraulic potential karstified systems such as the Jura, Alpine and Mediterranean karsts which have been the model for the establishment of multi-criteria methods such as EPIK and its derivatives, the low plateaus of the Northeast of France have a low water potential and at first sight a very discreet karstification because the dissolution of limestone primarily occurs under a non-carbonate cover. Thus, the karstification of these plateaus is made possible when low permeability facies (marly limestone, marl or clay) are overlying limestone formations. This cover allows the concentration of surface runoff to

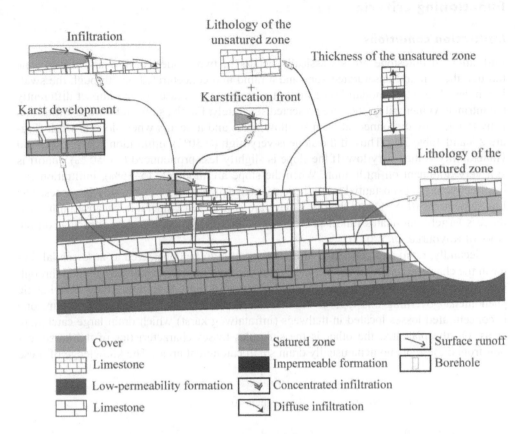

Figure 2.2 The aquifer characteristics and the choice of the criteria

karstic losses and a supply of the underlying limestones (Gamez, 1995) resulting in a crypto-corrosion and the appearance of an active karst (Figure 2.2). These preferential dissolution areas correspond to karstification fronts (Jaillet, 2000; Devos, 2010). In the major lime-stone valleys, karstification also occurs when a river coming from less permeable formations passes through the limestone (allogenic system). Other features differ to traditional karst aquifers; in the low bottom plateaus of the Northeast of France, the upper part of the aquifer infiltration zone is usually marked by low permeability layers that inhibit the existence of a true epikarst (Fister, 2012). The choice of criteria is guided both by the aquifer physical characteristics and by the dimension of the study area (8130 km²) which prevents to resort to field investigations. Criteria such as the nature and thickness of the soil can be taken into account in the method. The mapping only uses available data from national (BDLISA for geology, BDALTI for topography and slopes) and regional databases (BDIKARE for karst morphologies and tracer tests).

These features govern the choice of five criteria (Figure 2.2); two criteria are primarily interested in the flow dynamic implied by exokarst and endokarst. The other three criteria are interested in the structural characteristics of the aquifer i.e. the lithology and the thickness of the unsaturated zone and the lithology of the saturated zone.

Functioning criteria

Infiltration conditions

Infiltration in limestone areas classically occurs in two modes: a slow and diffuse one through the soil and unsaturated zone and a rapid and concentrated one through the swallow holes. These two modalities of infiltration refer to processes which impact differently the intrinsic vulnerability, they are presented separately. Firstly, slow and diffuse infiltration is hydrologically disconnected from swallow holes and it occurs when slope is weak (Figure 2.3 and Table 2.1). Thus, if the slope is very high (> 50%), infiltration is negligible and vulnerability is then very low. If the slope is slightly less pronounced (15–50 %), runoff is still very dominant on infiltration. When the slope are moderate (5–15%), infiltration can be effective during substantial rainfall events, consequently the vulnerability increases. The low slope areas (0–5 %) and the dry valleys are considered vulnerable because infiltration process largely dominates runoff process. This classification is based in particular on the work of Kavouri *et al.* (2011).

Secondly, rapid and concentrated infiltration should be considered as an essential factor in the characterisation of the intrinsic vulnerability because surface flows travel through large underground voids and directly join the saturated zone. Within the study area, karstic phenomena governing this type of infiltration take two forms. In the valley bottom, one is concentrated losses located in thalwegs (infratalweg karst) which drain large catchment areas. On the plateaux, the other is the sinkholes-losses characteristics of the karstification fronts; these phenomena usually drain small catchment areas. The knowledge of these

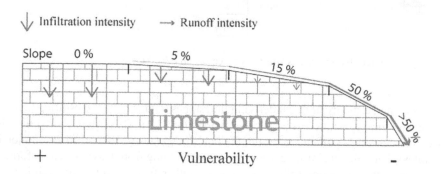

Figure 2.3 Slope and diffuse infiltration conditions

Table 2.1 Index of infiltration conditions

Index of infiltration conditions	
10	Slope higher than 50 %: runoff, no infiltration
11	High slope (15–50 %): mainly runoff, insignificant infiltration
12	High slope (15–50 %): mainly runoff, insignificant infiltration
13	Low slope (0–5 %): diffuse infiltration
14	Catchments areas of stream losses: concentred infiltration

morphologies is based on IKARE database that counts, locates, and characterises the different phenomena of the Ladinian, Bathonian-Bajocian and Oxfordian plateaus (François *et al.*, 2011). This database relies both on bibliographic work and field investigations carried out over the last fifty years by spelunkers and academics. It lists 638 phenomena whose catchments represent a combined surface area estimated at 1,883 km². The catchment areas of stream losses are assigned by a maximum vulnerability criterion.

Classes of slope were estimated via a DTM (BDALTI; DTM at 25m resolution). Associated with the infiltration phenomena, they allow to propose vulnerability classes of the infiltration criterion (Table 2.1 and Figure 2.7A).

Endokarstification

The hydrodynamic functioning of a limestone aquifer is related to its degree of karstification. One of the most commonly used methods to identify the degree of karstification is to use tracer tests which rapidly provide indications about the underground functioning. Within the study area, 175 injection campaigns were conducted; some areas concentrate almost all of the tracer test campaigns (only a few hundred km²), while the large majority of limestone outcrops have not been investigated. The tracer test information cannot constitute a sufficient information to understand the endokarstification of the different plateaus. So the choice was made to use in addition to tracer tests surface hydrological data. These data are based on basin hydrological yields during low water periods. During these periods, surface flow is only influenced by the aquifers; referring to discharges allows to assess underground contributions (Figure 2.4B).

The discharge used here is the mean monthly annual minimum flow recorded over a period of five years (MMAM 1/5) (AERM, 1998, 1999, 2000). For a subcatchment the yield is calculated, by the difference between the MMAM 1/5 of the downstream subcatchment and the MMAM 1/5 of the upstream subcatchment and then divided by the surface of the subcatchment under consideration (Figure 2.4A). Only subcatchments that completely or partially drain the three limestone plateaus were selected.

By superimposing the tracer tests on the hydrological yields map, there are clear connections between subcatchment marked by exceptional yields (negative or null, or extremely

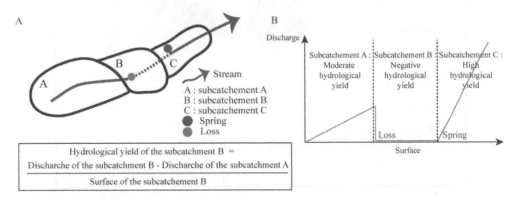

Figure 2.4 Calculation of the hydrological yields (A), impact of karstic morphologies on stream discharge (B)

strong) and sectors for which the tracer tests show rapid groundwater flow (Figure 2.5). Thus, in these sectors (Meuse system, for example, Figure 2.6), negative or null yields characterise stream losses while extremely high yields evoke concentrated underground contributions that are realized through karst springs. In such systems endokarstification is presumed important or very important. However there are sectors where this correspondence is not simply checked. In case of the Pays-Haut, tracer tests showed fast velocities transit but the hydrological yields can be considered as mean; disturbed by the underground extraction of iron ore, the sector is characterised by mine's dewatering that impact the surface flow and consequently the hydrological yields.

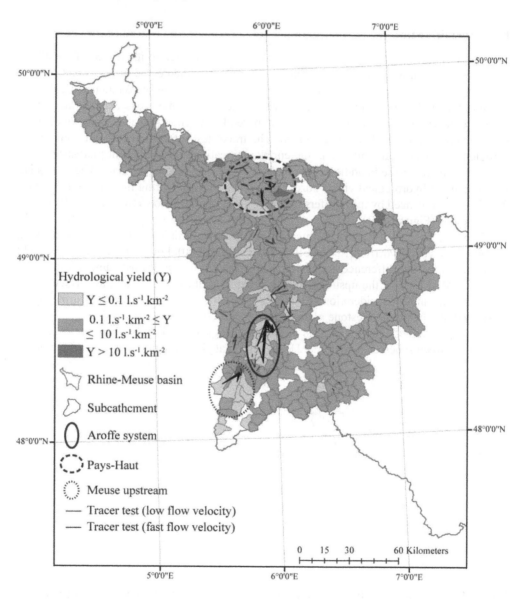

Figure 2.5 Hydrological yields and tracer tests of the Rhine-Meuse basin

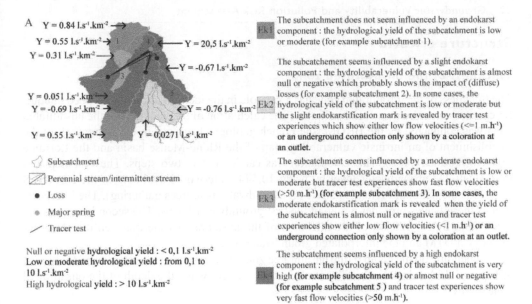

A

$Y = 0.84\ l.s^{-1}.km^{-2}$

$Y = 0.55\ l.s^{-1}.km^{-2}$

$Y = 0.31\ l.s^{-1}.km^{-2}$

$Y = 20.5\ l.s^{-1}.km^{-2}$

$Y = -0.67\ l.s^{-1}.km^{-2}$

$Y = 0.051\ l.s^{-1}.km^{-2}$

$Y = -0.69\ l.s^{-1}.km^{-2}$

$Y = -0.76\ l.s^{-1}.km^{-2}$

$Y = 0.55\ l.s^{-1}.km^{-2}$

$Y = 0.0271\ l.s^{-1}.km^{-2}$

Ek1 The subcatchment does not seem influenced by an endokarst component : the hydrological yield of the subcatchment is low or moderate (for example subcatchment 1).

Ek2 The subcatchment seems influenced by a slight endokarst component : the hydrological yield of the subcatchment is almost null or negative which probably shows the impact of (diffuse) losses (for example subcatchment 2). In some cases, the hydrological yield of the subcatchment is low or moderate but the slight endokarstification mark is revealed by tracer test experiences which show either low flow velocities (<=1 m.h⁻¹) or an underground connection only shown by a coloration at an outlet.

Ek3 The subcatchment seems influenced by a moderate endokarst component : the hydrological yield of the subcatchment is low or moderate but tracer test experiences show fast flow velocities (>50 m.h⁻¹) (for example subcatchment 3). In some cases, the moderate endokarstification mark is revealed when the yield of the subcatchment is almost null or negative and tracer test experiences show either low flow velocities (<1 m.h⁻¹) or an underground connection only shown by a coloration at an outlet.

Ek4 The subcatchment seems influenced by a high endokarst component : the hydrological yield of the subcatchment is very high (for example subcatchment 4) or almost null or negative (for example subcatchment 5) and tracer test experiences show very fast flow velocities (>50 m.h⁻¹).

◇ Subcatchment

▨ Perennial stream/intermittent stream

● Loss

● Major spring

╱ Tracer test

Null or negative **hydrological yield** : < 0,1 l.s⁻¹.km⁻²
Low or moderate hydrological yield : from 0,1 to 10 l.s⁻¹.km⁻²
High hydrological **yield** : > 10 l.s⁻¹.km⁻²

Note : The tracer test experiences made in the Rhine-Meuse basin show either very fast flow velocities (>50 m.h⁻¹) or very low flow velocties (<1 m.h⁻¹). There are no experiences that show intermediate flow velocities [1 m.h⁻¹ - 50 m.h⁻¹].

B

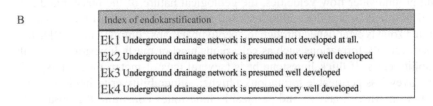

Index of endokarstification
Ek1 Underground drainage network is presumed not developed at all.
Ek2 Underground drainage network is presumed not very well developed
Ek3 Underground drainage network is presumed well developed
Ek4 Underground drainage network is presumed very well developed

Figure 2.6 Using hydrological yields and tracer test to assess Endokarstification (A, example with the Meuse upstream); index of this criteria (B)

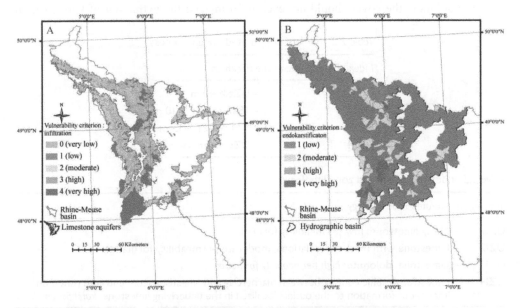

Figure 2.7 The functioning criteria maps: infiltration (A) and endokarstification (B)

Structure criteria

Thickness of the unsaturated zone

The thickness of the unsaturated zone has a key role in the identification of the speed of aquifer contamination by a potential pollution. The calculation of the thickness of the unsaturated zone was conducted by the BRGM (the French geological survey) in the framework of the establishment of an intrinsic vulnerability map of the Rhine-Meuse basin and the Lorraine region (Mardhel, 2010). This calculation was carried out in two steps. The first step was to evaluate the groundwater level by using 10,544 waterworks available in the BDADES (ADES is the national French database on groundwater resources gathering). The interpolation of these values provides a map of the first groundwater levels. The second step was the calculation by subtraction of the thickness of the unsaturated zone between the elevation given by a DTM and the interpolated groundwater levels.

Depending on the depth of the unsaturated zone, four criteria were established (Table 2.2 & Figure 2.10A); the vulnerability is obviously stronger when the depth of the unsaturated zone is low.

Lithology of the unsaturated zone

By influencing groundwater flow velocities, the geological nature of the unsaturated zone formations, b determines in part the infiltration conditions. The lithology of the formations of the unsaturated zone is established through the geological maps compiled in the BDLISA. Low matrix permeability formations such as calcareous marl or marly limestones form the lower vulnerability class. If thick limestone beds (several tens of meters) are interspersed with marly or clayey less permeable formations a higher vulnerability criterion is assigned. The configuration in which massive and permeable limestones constitute the unsaturated zone is assigned by a high vulnerability index (Table 2.3 & Figure 2.8).

The strongest vulnerability index for the unsaturated zone lithology criterion is formed by karstification fronts which are active dissolution areas conducive to infiltration. For an active dissolution, the cover should not exceed 20 meters; the extraction of karstification

Table 2.2 Index of unsaturated zone thickness

Index of unsaturated zone thickness	
UZtl	– UZt > 40 m
UZt2	– 15 m < UZt ≤ 40 m
UZt3	– 5 m < UZt ≤ 15 m
UZt4	– UZt ≤ 5 m

Table 2.3 Index of unsaturated zone lithology

Index of unsaturated zone lithology	
UZl1	– Marly limestone: low permeability formation
UZl2	– Limestone with marl intercalations: moderate permeability formation
UZl3	– Limestone, dolomite: *high permeability formation*
UZl4	– Lithostratographic contact between marne/clay and limestone, concentrated runoff on impervious formation to the outlets localised in the underlying limestone: *karstification front*

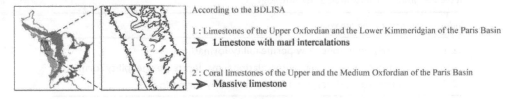

According to the BDLISA

1 : Limestones of the Upper Oxfordian and the Lower Kimmeridgian of the Paris Basin
⟶ Limestone with marl intercalations

2 : Coral limestones of the Upper and the Medium Oxfordian of the Paris Basin
⟶ Massive limestone

Figure 2.8 Identification of the lithology of the unsaturated zone

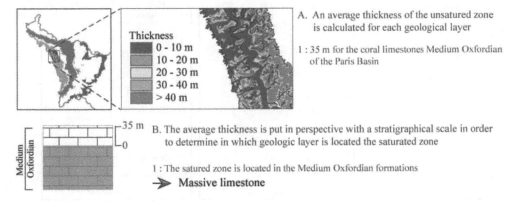

A. An average thickness of the unsaturated zone
   is calculated for each geological layer

1 : 35 m for the coral limestones Medium Oxfordian
   of the Paris Basin

B. The average thickness is put in perspective with a stratigraphical scale in order
   to determine in which geologic layer is located the saturated zone

1 : The saturated zone is located in the Medium Oxfordian formations
⟶ Massive limestone

Figure 2.9 Identification of the lithology of the saturated zone

fronts is therefore to identify the marly/clayey outcrops overlying limestone plateaus and for which the thickness does not exceed the 20 meters threshold. From a methodological point of view, the karstification fronts are individualised as follows. As a first step, lithological cover/limestone contrasts are identified; marly/clayey outcrops which sporadically cover the limestone plateaus are identified by referring to the BDLISA. The low permeability outcrop elevation is assessed by using a DTM. To individualise the 20 m threshold, it is necessary to know the roof elevation of the underlying limestone. This altitude is locally known thanks to boreholes. These borehole points, listed by the BRGM, provide information on the elevation of the different formations encountered; thus the altitude of the cover/limestone limits is locally known. These points are sample sites which are then used as a basis for interpolation and establishment of an isobaths map. The limestone roof elevation is spatialized via the inverse distance method. The difference between this extrapolated elevation and the elevation of the ground surface corresponds to the hypothetical cover thickness. The karstification front is individualised by keeping only values below 20 meters. On the study area, three karstification fronts are identified: on the Ladinian plateau a karstification front is developing between Keuper marl and Ladinian limestone, on the Bathonian-Bajocian plateau a karstification front is established between the marl of Gravelotte and the Jaumont limestone, and on the Oxfordian plateau a karstification front is developed between the upper Oxfordian marly limestones and the middle Oxfordian limestones.

Lithology of the saturated zone

The lithology of the saturated zone cannot be assessed directly. To identify which formations can be considered as a groundwater reservoir, data from the BDLISA and the thickness of the unsaturated zone are coupled according to the methodology proposed in Figure 2.9.

Table 2.4 Index of saturated zone lithology

Index of saturated zone lithology	
SZI1	– Marly limestone: low permeability formation
SZI2	– Limestone with marl intercalations: moderate permeability formation
SZI3	– Limestone, dolomite: high permeability formation

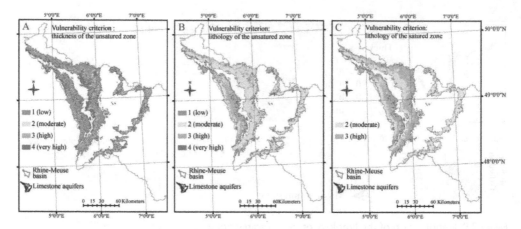

Figure 2.10 The structure criterion maps: thickness and lithology of the unsaturated zone (respectively A and B), lithology of the saturated zone (C)

The lithology of the saturated zone criterion is indexed in the same way as the lithology of the saturated zone criterion (Table 2.4 & Figure 2.10C), with the exception of karstification fronts.

Intrinsic vulnerability map & discussion

The calculation of the intrinsic vulnerability map of the limestone aquifers is based on a weighted combination of the five criteria according to the following equation: Intrinsic vulnerability = a*infiltration+ b*endokarstification+ c*thickness of the unsaturated zone+ d*lithology of the unsaturated zone+ e*lithology of the saturated zone (where a, b, c, d, e correspond to weights).

Three calculation principles are followed. (1) The sum of the weight equal to 1. (2) If, for a pixel, the criterion of vulnerability "infiltration" is 4, i.e. it corresponds to a catchment area of a stream loss, the total intrinsic vulnerability to that pixel is equal to 4; then this criterion is not weighted by the other criteria. (3) The sum of the functioning criteria (infiltration and endokarstification) is between 50% and 60% of the total weight; these criteria involving karst phenomena are considered as very impactful on potential contamination. The sum of the structural criteria varies from 40% to 50%.

Each pixel value is between 0 and 4, then five classes of equidistant values are calculated (Table 2.5).

Table 2.5 Vulnerability class

Class	Vulnerability
[0–0,79]	Very low 0
[0,8–1,59]	Low 1
[1,6–2,39]	Moderate 2
[2,4–3,19]	High 3
[3,2–4]	Very high 4

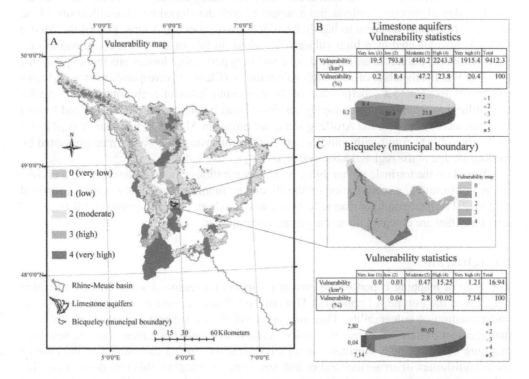

Figure 2.11 Intrinsic vulnerability map (A), vulnerability statistics (B) and vulnerability of the village of Bicqueley (C)

The intrinsic vulnerability map (Figure 2.11A) was calculated according to the following equation: Intrinsic vulnerability = 0.3*infiltration + 0.3*endokarstification + 0.15*thickness of the unsaturated zone + 0.15*lithology of the unsaturated zone + 0.1*lithology of the saturated zone.

Statistics were calculated in order to present the spatial extent of the different vulnerability classes, (Figure 2.11B). These figures show that these aquifers are generally vulnerable to pollution; only about 10% of the three limestone plateaus can be considered as non-vulnerable to pollution. In greater details, these low vulnerability areas, for the Oxfordian plateau, correspond to the hillslopes of the many tributary valleys of the Meuse valley. For the Bathonian-Bajocian and Ladinian plateaus, these sectors are mainly located at the edge of the plateaus (eastern end) where the slopes are steep and the thickness of the unsaturated zone is very important. It should also be noted that the northern part of the Bathonian-Bajocian plateau is generally not very vulnerable to pollution particularly because of a very low density of karst morphologies in this sector. Moderate vulnerability (class 2) largely dominates with 47 % of the total area. It reflects the non-karstic areas but for which the other criteria in particular the structural ones are affected by mean or high vulnerability index: permeable formations, shallow unsaturated zone, low slope. This figure (to which we can add the 8,5% that represent low vulnerability classes) reflects that a large part of these plateaus are disconnected from karst, which seems conform to field realities. High vulnerability class (class 3) represents 24% of the total area. This class mostly corresponds to the high vulnerability index of structural criteria and a supposed well developed endokarstification. These areas are frequently adjacent to highly karstified areas (class 4). A large part of the Meuse valley is also affected by a high vulnerability class; in this case it is the combination of a very low slope, a shallow unsaturated zone and very permeable formations that explain this ranking. Finally, the class of very high vulnerability (Class 5) corresponds to stream-losses catchment areas or to a superposition of the maximum index of each criterion (except for the infiltration criterion). These highly sensitive areas belong to well-karstified and known systems such as those of the Aroffe system, the upstream Meuse system, the "Pays-Haut"; so it is possible to empirically validate some of the intrinsic vulnerability map presented by the knowledge of the regional karst systems.

Based on the intrinsic vulnerability map, sectorisations can be undertaken on a chosen scale. For example, a sectorisation on the Bicqueley municipal boundary scale is proposed (Figure 2.11C), the resulting map reflects risk areas and can be used for the establishment of structures that are potential threats to groundwater quality.

Conclusion

This type of mapping is part of the current prospects for reasoned water management that materialize the European Framework Directive on Water because it contributes to a better understanding of vulnerability. This map is intended as a support to the water resources management of limestone areas that guides stakeholders and direct their choices: however, mapping could not be a substitute for an environmental impact assessment in the case of the establishment of an agricultural or industrial site. In addition, this tool doesn't provide a fine-scale characterisation of vulnerability, it is a support document, but it can in no way replace the hydrogeological work of protection areas delineation; a change of investigation scale is then essential.

Providing vulnerability maps at this scale obviously depends on the deficiencies in the databases that are used. A major problem for the map achievement on this scale is to have reliable and well-informed criteria. But it appeared that the criteria are not always well known, particularly those related to karst: the BDIKARE cannot be considered as completely exhaustive. The exokarst morphologies are not all known or well localised involving biases in the establishment of loss catchment. The results of the tracer test used to characterise the endokarstification criteria may also be subject to reservations; sometimes the information is only based on visual observations, in other cases the information are is complete and includes features such as the minimum transit speed. In the vast majority of the cases, the residence time distribution has not been established due to the lack of knowledge of the emergence discharges. We should also note that the data that quantify the structural criteria suffer from similar pitfalls; geological logs are sometimes inaccurate, geological delineations do not always correspond to field reality. . . . The unsatisfactory data quality leads us to admit that the proposed vulnerability cannot always be considered as perfectly faithful to field realities. Taking into account the spatial extent of the study area, it would hardly have been feasible to fill in these gaps. Still due to the size of the study area, some parameters that influence the vulnerability could not be taken into account such as the nature and thickness of soil or rock fracturing.

Also, in a second time a scale change will be set up. It consists in a zoom on catchment areas of capture works used for drinking water. The criteria used at the aquifer scale will be kept but assessed by field surveys. Firstly this approach will provide a fine quantization of the vulnerability of the followed water supply works and second these results will be compared to those obtained at the aquifer scale in the perspective of the validation of the small-scale maps.

Acknowledgements. This research has been started in October 2014 in partnership between the LOTERR laboratory, the Rhine-Meuse Water Agency and the Lorraine region. The authors would like to thank X. Marly and L. Vaute for their active contribution.

References

AERM (ed.) (1998) *Débits mensuels d'étiage et modules. Vol. 2: Bassin de la Moselle amont (Bassin de la Meurthe inclus)*. Agence de l'eau Rhin-Meuse, Moulins-lès-Metz, France.

AERM (ed.) (1999) *Débits mensuels d'étiage et modules. Vol. 4: Bassin de la Meuse*. Agence de l'eau Rhin-Meuse, Moulins-lès-Metz, France.

AERM (ed.) (2000) *Débits mensuels d'étiage et modules. Vol. 3: Bassin de la Moselle aval (Sarre inclus)*. Agence de l'eau Rhin-Meuse, Moulins-lès-Metz, France.

AERM (ed.) (2009) *SDAGE 2010–2015. Directive cadre européenne sur l'eau, tome 1*. Agence de l'eau Rhin-Meuse, Moulins-lès-Metz, France.

Davies, A.D., Long, A.J. & Wireman, M. (2002) KARSTIC: A sensitive method for carbonate aquifers in karst terrain. *Environmental Geology*, 42, 65–72.

Devos, A. (2010) *Les conditions d'écoulement des plateaux calcaires de l'Est de la France*, Volume 1. (303p). *Mémoire d'habilitation à diriger les recherches (HDR)*, Université de Champagne-Ardenne, Reims, France.

Dörfliger, N. & Zwahlen, F. (1998) *Groundwater Vulnerability Mapping in Karstic Region (EPIK), Practical Guide*. Swiss Agency for the Environment, Forests and Landscape (SAEFL), Berne.

Fister, V. (2012) *Dynamique des écoulements dans les aquifères calcaires de bas plateaux: de l'identification à la quantification des types de circulation. Exemple des formations triasiques et jurassiques dans le Nord-Est de la France*. Thèse de doctorat, Université de Lorraine, Metz, France.

François, D., Losson, B. & Harmand, D. (2011) *IKARE: inventaire des phénomènes karstiques et des écoulements en milieu calcaire. Base de données spéléo-karstologique du bassin Rhin-Meuse et des régions limitrophes*. CEGUM, Metz.

Gamez, P. (1995) *Hydrologie et karstologie du bassin du Loison (Woëvre septentrionale – Lorraine)*. Mosella, PUM, Metz, t. XXI, n° spécial annuel (1991). Thèse de doctorat, Université de Metz, France.

Goldscheider, N. (2005) Karst groundwater vulnerability mapping: Application of a new method in the Swabian Alb, Germany. *Hydrogeology Journal*, 13(4), 555–564.

Jaillet, S. (2000) *Un karst couvert de bas-plateau: le Barrois (Lorraine /Champagne, France). Structure-Fonctionnement-Évolution*. Thèse de doctorat, Université de Bordeaux III, France.

Kavouri, K., Plagnes, V., Tremoulet, J., Dörfliger, N., Rejiba, F. & Marchet, P. (2011) PaPRIKa: a method for estimating karst resource and source vulnerability: Application to the Ouysse karst system (southwest France). *Hydrogeology Journal*, 19, 339–353.

Mangin, A. (1975) *Contribution à l'étude hydrodynamique des aquifères karstiques.* Annales de Spéléologie, t.29 et t.30, fasc. 3, 4 et 1 Thèse d'Etat, pp. 282–332, 495–601 et 21–124, Université de Dijon, France.

Mardhel, V. (201 0) *Cartographie de la vulnérabilité intrinsèque simplifiée des eaux souterraines du bassin Rhin-Meuse et de la région Lorraine.* BRGM, rapport public-56539, Orléans, France.

Pételet-Giraud, E., Dörfliger, N. & Crochet, P. (2000) RISKE: méthode d'évaluation multicritère de la vulnérabilité des aquifères karstiques. Application aux systèmes des Fontanilles et Cent-Fonts (Hérault, Sud de la France). *Hydrogéologie*, 4, 71–88.

Pranville, J., Plagnes, V., Rejiba, F. & Trémoulet, J. (2008) Cartographie de la vulnérabilité sur la partie sud du causse de Gramat: application de la méthode RISKE 2. *Géologues*, 156, 44–48.

Chapter 3

Numerical simulations of aquifer vulnerability to methane gas leakage from decommissioned shale gas wells

N. Roy, J. Molson, J.-M. Lemieux,
D. Van Stempvoort & A. Nowamooz

1 Introduction

Leakage of natural gas along producing or decommissioned hydrocarbon wells is a recurrent problem in the oil and gas industry. Indeed, oil and gas field operations such as well stimulation and production can allow formation fluids and natural gas (primarily methane) to migrate from the production or intermediate zones to the subsurface, along preferential pathways such as poorly cemented casings and improperly sealed decommissioned wells (Dusseault et al., 2014; Nowamooz et al., 2015; Ryan, 2015). According to Bexte et al. (2008), 7 to 19% of the wells drilled between 2005 and 2007 in western Canada were affected by gas migration along the annulus while 9 to 28% showed gas leakage through the surface casing. In particular, gas migration along production wells was observed by Erno and Schmitz (1996) in the Lloydminster area in western Canada.

Flow rates from leaky wells can be enhanced by overpressures often associated with low permeability rocks, resulting in naturally high upward pressure gradients (Flewelling and Sharma, 2013). Nowamooz et al. (2013, 2015) simulated methane flow rates along the annulus of a shale gas production well for different cement permeabilities and showed that cement quality can also significantly affect the rate of upward gas migration. Thus, under suitable conditions, there is a significant risk that natural gas and brine saturated with methane could migrate along a well and penetrate shallow aquifers. The risk of contamination will depend on the vulnerability of the aquifer, including the capacity of the aquifer to attenuate methane through dispersion and natural biodegradation. Aquifer vulnerability to this type of risk has not yet been fully defined.

Once natural gas reaches a shallow aquifer, it can have significant impacts on groundwater chemistry. Kelly et al. (1985), for example, observed groundwater chemistry turnover following a gas well blowout in North Madison, Ohio, USA. Although the ingestion of methane is not toxic to human health, concentrations above its solubility can result in explosion and asphyxia hazards (Vidic et al., 2013). A threshold of 10 mg/L is used by the US Office of the Interior to identify wells at possible risk. Consequently, it is crucial to understand on one hand the behaviour of methane in the subsurface and on the other hand the natural attenuation processes that can mitigate methane concentrations in soils and aquifers and thus decrease aquifer vulnerability.

Methane oxidation is an important sink of methane and can occur in a variety of environments, including groundwater. Indeed, Van Stempvoort et al. (2005) found geochemical evidence of coupled methane oxidation and bacterial sulfate reduction near a leaky oil production well in a confined aquifer in western Canada. Their data provided strong evidence of active methane biodegradation, especially at the top of the aquifer.

Building on the previous studies of Van Stempvoort and Jaworski (1995), Van Stempvoort *et al.* (1996), Maathuis and Jaworski (1997) and Van Stempvoort *et al.* (2005), numerical simulations are here presented to evaluate the vulnerability of shallow confined aquifers to methane gas migration from deep geological formations. A sensitivity analysis is conducted to understand the effect of leakage duration and gas inflow rate, and to evaluate the efficiency of bacterially-driven sulfate reduction to attenuate methane concentrations. Some conclusions are drawn with respect to assessing aquifer vulnerability to methane contamination.

2   Conceptual model

The conceptual model is based loosely on the Lloydminster field site described by Van Stempvoort *et al.* (2005) which includes a homogeneous, isotropic and confined aquifer of constant thickness pierced by an abandoned hydrocarbon production well (Figure 3.1). The adopted conceptual model is quite common in glacial terrains in Canada and northern Europe where permeable sand, gravel or rock aquifers are overlain by low permeability glacial till or marine clay. Confined aquifers should also represent the most conservative (most vulnerable) scenario, compared to unconfined aquifers.

Gas entering at the base of the initially water-saturated aquifer is assumed to be only methane. Representative surface casing vent flows (SCVFs) ranging from 0.01 to 10 m³/day (measured at surface pressure and temperature conditions) for wells in western Canada were considered based on documented cases (Erno and Schmitz, 1996; Dusseault *et al.*, 2014; Nowamooz *et al.*, 2015). Gas flux along the well annulus was assumed to be constant in time and to stop abruptly after a given period.

As methane solubility in water is relatively low (assumed in the model as 110 mg/L at 25 m depth and 5°C), the mass flux of dissolved methane entering the aquifer was assumed

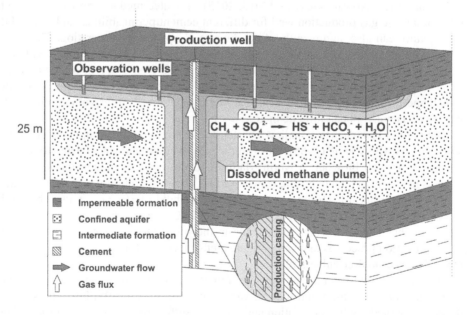

Figure 3.1 Conceptual model for the numerical simulations showing a confined aquifer intersected by a leaky production well.

Table 3.1 Physical and hydrodynamic parameters used in the numerical model

Parameter	Value
Hydraulic conductivity	6×10^{-5} m/s
Porosity	0.35
Hydraulic gradient	5×10^{-5}
Longitudinal dispersivity (α_L)	5 m
Transverse horizontal dispersivity (α_{TH})	2.5 m
Transverse vertical dispersivity (α_{TV})	0.01 m
CH_4 effective diffusion coefficient	10^{-9} m²/s

negligible with respect to the gas phase (Nowamooz *et al.*, 2013), and brine influx was assumed to not disturb the natural hydraulic gradient. Consequently, only the gas phase entering the aquifer was considered. Capillary pressures and gas phase residual saturation were set to zero since the aquifer is permeable and porous and it exhibits no resistance to the flow of gas. The overlying aquitard was assumed impermeable to water and gas.

When the methane concentration exceeds its solubility, a gas phase may appear and migrate upward by buoyancy via the annulus zone where it would accumulate at the top of the confined aquifer. The progressive dissolution of the methane gas phase into groundwater produces a dissolved methane plume that is progressively transported by advection and hydrodynamic dispersion. Methane consumption is presumed to not affect the magnitude of the gas phase dissolution rate. Finally, considering the low groundwater flow rate, methane dissolution into groundwater is assumed to be at equilibrium, thus will dissolve into the groundwater at its solubility concentration of 110 mg/L.

The mass of dissolved gas can be reduced by in-situ methane oxidation linked to bacterial sulfate reduction. The following bacterially-driven reaction was proposed to be effective near the production well at the Lloydminster site (Van Stempvoort *et al.*, 2005):

$$CH_4 + SO_4^{2-} \rightarrow HS^- + HCO_3^- + H_2O \tag{3.1}$$

Only limited information is available on the kinetic rate of this reaction in shallow aquifers. However, according to Knittel and Boetius (2009) the reaction rate depends largely on the availability of methane and sulfate concentrations. Thus, it is possible to infer a representative range of values for the maximum methane oxidation rate based on existing data published in the literature for other environmental contexts. We assume a maximum oxidation rate ranging from 1×10^{-5} to 1×10^{-3} kg/m³/day based on published data (Kosiur and Warford, 1979; Bussman *et al.*, 1999). Table 3.1 summarizes the physical and chemical parameters used in the model.

3   Simulation strategy

Transient multi-phase methane migration simulations were first launched with the DuMu[x] model (Flemisch *et al.*, 2011) to determine gas-phase methane saturations in the aquifer for a wide range of gas inflow rates and leakage durations, and assuming no methane degradation. At each time step, the resulting methane saturation distributions were coupled to the BIONAPL/3D model (Molson and Frind, 2018). The transient BIONAPL simulations accounted for natural background flow, equilibrium dissolution of methane into the flowing groundwater, advective-dispersive transport of the dissolved-phase methane plume, as well as anaerobic biodegradation for the reactive cases.

4   Domain discretisation, boundary conditions and leakage rates

A 3D structured grid, symmetrical about the vertical plane through the production well and oriented parallel to the groundwater flow (thus representing one-half of the entire domain), was used in both numerical models. The simulated aquifer domain is 250 m long, 50 m wide (being the half-width) and 25 m deep, and is discretised with $141 \times 63 \times 23$ ($=204,309$) elements in the respective directions. The production well is 0.2 m in diameter and crosses the entire thickness of the aquifer.

In the DuMu^x multi-phase model, the domain was assumed to be initially water-saturated. A fixed horizontal hydraulic gradient of 5×10^{-5} was initially imposed over the domain, using fixed water pressures on the left (inflow) and right (outflow) faces, which created a groundwater velocity of 7.4×10^{-4} m/day (consistent with the field data). A fixed gas saturation was imposed on the left face while in order to allow gas to leave the domain, an gas outflow boundary condition was set on the right face. Finally, no-flow boundary conditions were assigned on the remaining faces.

Initial and boundary conditions assigned in the BIONAPL/3D model for groundwater flow and dissolved gas transport were consistent with the DuMu^x simulations. An identical hydraulic gradient was imposed across the domain using Dirichlet conditions (fixed heads) at each end. No-flow boundary conditions were assigned on the remaining faces. The initial dissolved methane and sulfate concentrations were set to 0 and 400 mg/L, respectively, for the base case simulation throughout the domain (with sulfate only used in the reactive simulations – Cases 7–10). A fixed sulfate concentration on the left inflow face (equal to background) and zero-gradient conditions for sulfate and methane were set on the remaining faces. In the reactive cases, methane biodegradation was assumed governed by dual-Monod type kinetics (Molson and Frind, 2018; Schirmer et al., 2000), assuming an initial microbial concentration of 1.5×10^{-3} kg/m^3, half-utilization constants of 0.055 and 0.2 kg/m^3 for methane and sulfate, respectively (one-half their maximum concentrations), and a microbial yield coefficient of 0.03. Maximum degradation rates are given below.

Impacts of gas migration were assessed through four gas inflow rates of 1×10^{-7}, 1×10^{-6}, 1×10^{-5}, and 1×10^{-4} kg/s, covering approximately the above-mentioned range between 0.01 and 10 m^3/day (7.7×10^{-8} to 7.7×10^{-5} kg/s) (Erno and Schmitz, 1996; Dusseault et al., 2014; Nowamooz et al., 2015). These rates were applied for an arbitrary duration of two years at the base of the aquifer. As a base case condition, only the processes of gas dissolution, advection and dispersion were considered while methane biodegradation was neglected. Time steps varying from 0.01–1 days were applied.

5   Results and discussion

Simulated evolution of the dissolved methane from the (non-reactive) base case, with the two highest leakage rates (1×10^{-5} and 1×10^{-4} kg/s; 1.3 and 13 m^3/day, respectively) is shown in Figure 3.2 (lower gas inflow rates of 1×10^{-7} and 1×10^{-6} kg/s did not result in

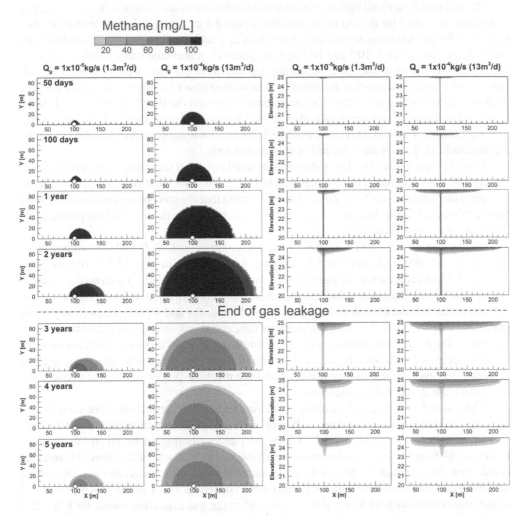

Figure 3.2 Simulated dissolved methane plume evolution from the conservative base case with no degradation: plan views at the top of the aquifer (left two columns) and longitudinal cross-sections through the well (right two columns). Methane concentrations are shown for two different methane gas inflow rates, at selected times over 5 years. The leakage ends after 2 years.

significant gas plumes at the top of the aquifer). In each case, the gas phase migrated upward due to buoyancy and accumulated at the top of the aquifer, with the plume growing over time. An increase in the gas inflow rate of one order of magnitude resulted in a methane plume area approximately 3 times greater after 2 years, with a somewhat greater thickness. Due to the assumed slow background flow velocity, the pool of dissolved methane shows only limited downgradient migration from the leaky well. Once the gas inflow stopped (after 2 years), the dissolved gas concentration progressively decreased due to dispersion but concentrations of 50 mg/L remained after 5 years.

Breakthrough (arrival) curves of methane concentrations are shown in Figure 3.3a, b, c at a monitoring well (at the top of the aquifer) situated 5 m downgradient from the production well, for the ten considered cases. For Cases 2, 3 and 4 (all without degradation, with gas leakage rates of 10^{-6}, 10^{-5} and 10^{-4} kg/s, respectively), methane concentrations increase rapidly during the first few days, indicating gas arrival at the top of the aquifer. The sudden end of the gas leakage was followed by a rapid decrease of methane concentrations but with significant tailing over time. Concentrations exceeded the threshold of 10 mg/L in all cases and remained above the threshold over the 25-year time period for leakage rates above 10^{-5} kg/s (Cases 3 and 4). However, in Case 2, the methane concentration decreased below the threshold 16.7 years after the end of the gas leakage.

Case 1 (Figure 3.3a, 10^{-7} kg/s) showed a delayed arrival of the dissolved gas of approximately 100 days compared to other cases. Moreover, the associated concentration peak did not reach the methane solubility limit, clearly showing that the gas phase spread significantly less in this case. Finally, the gas concentration decreased below the threshold approximately 2.7 years after the end of the gas leakage.

As showed by Erno and Schmitz (1996), Dusseault *et al.* (2014) and Nowamooz *et al.* (2015), most leaky wells have SCVFs ranging from 1 to 10 m³/day (7.7×10^{-6} to 7.7×10^{-5} kg/s). Thus a representative gas inflow rate of 10^{-5} kg/s (1.3 m³/day) with no degradation was set for all subsequent simulations. Based on these conditions, two additional conservative cases (Case 5 and 6) with leakage durations of 6 months and 4 years, respectively, were simulated (Figure 3.3b). As expected, methane concentrations were particularly sensitive to the gas leakage duration. Indeed, with the shortest duration (6 months), methane concentrations decreased to below the threshold concentration 17.1 years after the gas migration stopped. However, Case 6 (4-year duration) did not meet the proposed threshold concentration within the 25-year simulation time frame.

The natural processes of methane oxidation and sulfate reduction were then evaluated for the base case using three different methane oxidation rates of 10^{-5}, 10^{-4}, and 10^{-3} kg/m³/day (Figure 3.3c and 3d – Cases 7–9) and assuming an initial sulfate concentration of 400 mg/L. In these conceptual simulations, by-products of reaction (1) were not considered. As expected, the higher the maximum oxidation rate, the greater the decrease in dissolved methane with time. Indeed, the threshold concentration of 10 mg/L is now reached in the observation well 18.8, 3.5 and 2.5 years after the gas migration ended for Case 7, 8 and 9, respectively. This suggested that dissolved gas concentrations were very sensitive to the maximum rate of methane consumption. However, the difference in methane concentration between cases became less important as the maximum oxidation rate increased. Further details can be found in Roy *et al.* (2016).

In southern Quebec (Canada), natural background sulfate concentrations in aquifers situated in the prospective shale gas areas between Quebec City and Montreal are typically about 40 mg/L (Blanchette, 2006; Carrier *et al.*, 2013; Larocque *et al.*, 2013). In order to evaluate aquifer vulnerability to deep gas migration, this average value was applied as a new initial sulfate concentration with an average methane oxidation rate of 10^{-5} kg/m³/day (Figure 3.3c and 3.3d – Case 10). The decrease in methane concentration was here less pronounced than for Case 8 (with the same oxidation rate but with 400 mg/L SO$_4$) and the threshold concentration was reached only 22.8 years after the gas migration ended. This highlights the vulnerability of these aquifers to gas migration. As shown in Figure 3.3d, sulfate concentrations decreased abruptly and were negligible after about 1 year, resulting in a strong decrease in the methane oxidation rate.

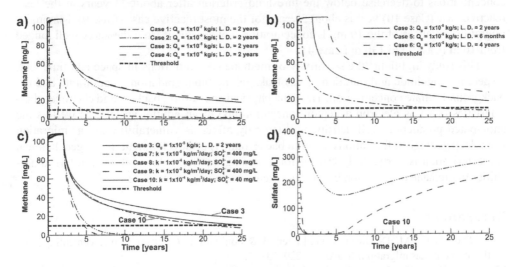

Figure 3.3 Breakthrough curves of methane (a, b, c) and sulfate concentrations (d) for an observation well 5 m downgradient from the leaky production well, for Cases 1 to 10. Sulfate concentrations shown for reactive cases (Cases 7–10). L.D. = leakage duration.

As expected, the higher the maximum oxidation rate, the stronger the decrease in sulfate concentration with time (Case 7, 8, 9). In the most reactive case (Case 9), the sulfate concentrations decreased below 1 mg/L about 1.5 years after the beginning of the gas leakage. Interestingly, the decrease in sulfate concentration was less pronounced for Case 10 (40 mg/L SO_4) than for Case 8 (400 mg/L SO_4), reflecting the lower availability of sulfate in the medium. In each case, the sulfate concentrations tended to start recovering to background levels once the methane concentration began to decrease.

6 Conclusions

Three-dimensional numerical simulations of a field-based conceptual model representing dissolution of methane gas and downgradient reactive transport of methane from a leaky well into a confined aquifer have been presented. Although shallow confined aquifers can be vulnerable to this type of contamination, the simulations suggest that under some conditions ($[SO_4] \geq 400$ mg/L and degradation rates $\geq 10^{-4}$ kg/m³/d), natural biodegradation can significantly attenuate dissolved methane concentrations. The model is simplified and case-specific, thus the results should be used only as an example. For the assumed field case, a full model calibration would be required.

The simulations show that under this confined aquifer scenario, the gas phase plume pooled at the top of the aquifer due to its low density. Significant variability in the gas phase plume dimension was observed with respect to the gas inflow rate and leakage duration. Numerical simulations of unconfined conditions, currently in progress, suggest that unconfined aquifers would be less vulnerable to methane contamination.

All simulation cases showed peak methane concentrations above the threshold concentration of 10 mg/L but dispersion and bacterially-driven methane oxidation allowed concentrations to decrease below the threshold criterion after about 23 years in the less-reactive case (Case 10) versus about 3 years for the most reactive case (Case 9). Maximum migration distances of the 10 mg/L concentration downgradient from the leaky well ranged from 10 m (Case 1) to 130 m (Case 4).

This study highlights the sensitivity of methane concentrations in space and time with respect to magnitude and duration of gas leakage, the biodegradation reaction rate and the background sulfate concentration. Thus, the physical and natural background chemical characteristics (such as electron acceptor concentrations) of an aquifer system, as well as the anticipated production well density, will strongly affect its vulnerability to gas migration from deep geological formations. Natural background (pre-development) hydrogeochemical characterisation is particularly critical for assessing aquifer vulnerability. Aquifer vulnerability to methane degradation by-products remains an important issue.

References

Bexte, D.C., Willis, M., De Burjin, G.G., Eitzen, B. & Fouillard, E. (2008) Improved cementing practice prevents gas migration. *World Oil*, 229, 73–75.

Blanchette, D. (2006) *Caractérisation géochimique des eaux souterraines du bassin versant de la rivière Châteauguay, Québec, Canada.* (147p). Mémoire, Université du Québec, INRS, Québec.

Bussman, I., Dando, P.R., Niven, S.J. & Suess, E. (1999) Groundwater seepage in the marine environment: Role for mass flux and bacterial activity. *Marine Ecology Progress Series*, 178(1), 169–177.

Carrier, M.-A., Lefebvre, R., Rivard, C., Parent, M., Ballard, J.-M., Benoit, N., Vigneault, H., Beaudry, C., Malet, X., Laurencelle, M., Gosselin, J.-S., Ladevèze, P., Thériault, R., Beaudin, I., Michaud, A., Pugin, A., Morin, R., Crow, H., Gloaguen, E., Bleser, J., Martin, A. & Lavoie, D. (2013) *Portrait des ressources en eau souterraine en Montérégie Est, Québec, Canada.* [Online] (268pp). PACES Final Report INRS R-1433, (in French). Available from: www.environnement.gouv.qc.ca/_PACES/rapports-projets/MonteregieEst/MON-scientif-INRS-201306.pdf [Accessed 26 May 2019].

Dusseault, M., Jackson, R.E. & MacDonald, D. (2014) *Towards a Road Map for Mitigating the Rates and Occurrences of Long-Term Wellbore Leakage.* (69p). [Online]. University of Waterloo & Geofirma Engineering Ltd. Available from: http://geofirma.com/wp-content/uploads/2015/05/lwp-final-report_compressed.pdf [Accessed 26 May 2019].

Erno, B. & Schmitz, R. (1996) Measurements of soil gas migration around oil and gas wells in the Lloydminster area. *Journal of Canadian Petroleum Technology*. [Online], 35(7), 37–46. doi:10.2118/96-07-05 [Accessed 26 May 2019].

Flemisch, B., Darcis, M., Ebertseder, K., Faigle, B., Helmig, R., Lauser, A., Mosthaft, K., Müthing, S., Nuske, P., Tatomir, A. & Wolff, M. (2011) DuMux: DUNE for multi-{phase, component, scale, physics, . . .}, flow and transport in porous media. *Advances in Water Resources*, 34(9), 1102–1112.

Flewelling, S.A. & Sharma, M. (2013) Constraints on upward migration of hydraulic fracturing fluid and brine. *Ground Water*. [Online], 52(1), 9–19. doi:10.1111/gwat.12095 [Accessed 26 May 2019].

Kelly, W., Matisoff, G. & Fisher, J. (1985) The effects of a gas well blow-out on groundwater chemistry. *Environmental Geology and Water Sciences*, 7(4), 205–213.

Knittel, K. & Boetius, A. (2009) Anaerobic oxidation of methane: Progress with an unknown process. *Annual Review of Microbiology*. [Online], 63, 311–334. doi:10.1146/annurev.micro.61.080706.093130 [Accessed 26 May 2019].

Kosiur, D.R. & Warford, A.L. (1979) Methane production and oxidation in Santa Barbara Basin sediments. *Estuarine and Coastal Marine Science*. [Online], 8(4), 379–385. doi:10.1016/0302-3524(79)90054-9 [Accessed 26 May 2019].

Larocque, M., Gagné, S., Tremblay, L. & Meyzonnat, G. (2013) *Projet de connaissance des eaux souterraines du bassin versant de la rivière Bécancour et de la MRC de Bécancour – Rapport final.*

[Online] (219p). Rapport déposé au Ministère du Développement durable, de l'Environnement, de la Faune et des Parcs. Available from: https://rqes.ca/paces-becancour/ [Accessed 26 May 2019].

Maathuis, H. & Jaworski, E. (1997) *Migration of Methane into Groundwater from Leaking Production Wells Near Lloydminster: Summary Report 1994–1996.* (22p). Saskatchewan Research Council, Saskatoon, Saskatchewan, Canada.

Molson, J.W. & Frind, E.O. (2018) *BIONAPL/3D User Guide, a 3D Coupled Flow and Multi-Component NAPL Dissolution and Reactive Transport Model.* Université Laval, Québec City, Quebec & University of Waterloo, Waterloo, Ontario, Canada (Unpublished).

Nowamooz, A., Lemieux, J.-M. & Therrien, R. (2013) *Modélisation numérique de la migration du méthane dans les Basses-Terres du Saint-Laurent. Rapport final*, (in French). (115p). Québec Ministry of Environment, Québec.

Nowamooz, A., Lemieux, J.-M., Molson, J.W. & Therrien, R. (2015) Numerical investigation of methane and formation fluid leakage along the casing of a decommissioned shale-gas well. *Water Resources Research.* [Online], (51). Available from: http://dx.doi.org/10.1002/2014WR016146 [Accessed 26 May 2019].

Roy, N., Molson, J., Lemieux, J.-M., van Stempvoort, D. & Nowamooz, A. (2016) Three-dimensional numerical simulations of methane gas migration from decommissioned hydrocarbon production wells into shallow aquifers. *Water Resources Research.* [Online]. Available from: http://dx.doi.org/10.1002/2016WR018686 [Accessed 26 May 2019].

Ryan, M.C. (ed.) (2015) *Subsurface Impacts of Hydraulic Fracturing: Contamination, Seismic Sensitivity, and Groundwater Use and Demand Management.* Canadian Water Network HF-KI Subsurface Impacts Report. [Online]. Available from: http://cwn-rce.ca/report/2015-water-and-hydraulic-fracturing-report/ [Accessed 26 May 2019].

Schirmer, M., Molson, J.W., Frind, E.O. & Barker, J.F. (2000) Biodegradation modelling of a dissolved gasoline plume applying independent lab and field parameters. *Journal of Contaminant Hydrology*, 46(4), 339–374.

Van Stempvoort, D.R. & Jaworski, E. (1995) *Migration of Methane into Groundwater from Leaking Production Wells Near Lloydminster: Publication 1995–0001.* (46p). Canadian Association of Petroleum Producers, Calgary, Canada.

Van Stempvoort, D.R., Jaworski, E. & Rieser, M. (1996) *Migration of Methane into Groundwater from Leaking Production Wells Near Lloydminster: Report Phase 2: Publication 1996–0003.* (73p). Canadian Association of Petroleum Producers, Calgary, Canada.

Van Stempvoort, D.R., Maathuis, H., Jaworski, E., Mayer, B. & Rich, K. (2005) Oxidation of fugitive methane in ground water linked to bacterial sulfate reduction. *Ground Water.* [Online], 43(2), 187–199. doi:10.1111/j.1745-6584.2005.0005.x [Accessed 26 May 2019].

Vidic, R.D., Brantley, S.L., Vandenbossche, J.M., Yoxtheimer, D. & Abad, J.D. (2013) Impact of shale gas development on regional water quality. *Science.* [Online], (340). doi:10.1126/science.1235009 [Accessed 26 May 2019].

Chapter 4

Sanitary protection zoning of groundwater sources in unconsolidated sediments based on a Time-Dependent Model

V. Živanović, I. Jemcov, V. Dragišić & N. Atanackovic

I   Introduction

The basic task of groundwater source protection is to prevent contaminants from reaching the potential users. Delineation of source protection zones is a complex task, which relies on several factors such as the geological setting and the hydrogeological conditions, recharge-discharge relations, level of vulnerability, exploitation regime and contaminant properties. (European Commission, 2007).

Sanitary protection zoning methods for groundwater sources developed in intergranular aquifers are mainly based on the horizontal groundwater travel time from a given point to the source. Low groundwater velocities usually result in a high attenuation capacity, and vice-versa (Van Waegeningh, 1985). Sanitary zones are mainly defined using prescribed travel times (e.g. 50 days and 200 days). These travel times are selected to reduce the potential pathogens (50 days) or other harmful contaminants (200 days) to an acceptable level. Numerical models are widely used since different hydraulic conditions can be simulated, particularly in the case of complex groundwater sources (Kresic, 2007).

When an approach based on the horizontal travel time is applied, the main obstacle is that it tends to disregard the protecting role of the overlying strata (Parise *et al.*, 2015). In some cases, the overlaying deposits can completely attenuate a contaminant released from the ground surface. The protecting role of these layers is increasingly being addressed using vulnerability assessment methods (Zwahlen, 2004). The recently developed method – TDM for groundwater source vulnerability assessment (Živanović *et al.*, 2016) considers both horizontal and vertical transport components. This method has been designed to assess source vulnerability based on surface water and groundwater travel times. TDM was initially created for karst groundwater sources but is also applicable to other types of aquifers.

2   Applied method

The time-dependent model (TDM) for groundwater source protection zoning (Živanović *et al.*, 2016) considers three main components (Figure 4.1): surface water travel time (t_s) for areas in the source catchment where surface runoff is predominant; vertical travel time (t_v) through the unsaturated zone, or through the overlying low-permeability strata in the case of a confined aquifer; and horizontal travel time (t_h) through the saturated zone to the intake groundwater sources.

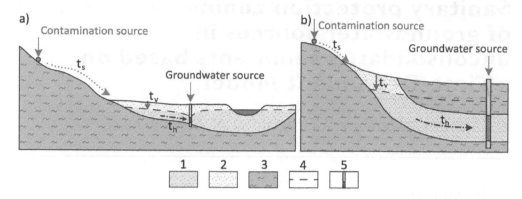

Figure 4.1 Conceptual TDM for intergranular groundwater source vulnerability assessment a) alluvial unconfined aquifer b) deep confined Neogene aquifer 1. Sand and gravel (saturated zone); 2. Sand and gravel (unsaturated zone); 3. Low-permeability rocks; 4. Groundwater/piezometric level; 5. Water supply well.

The total travel time of surface water and groundwater (t_{tot}), as well as of a contaminant from the ground surface to the source, can be estimated as follows:

$$t_{tot} = t_s + t_v + t_h \tag{4.1}$$

where:

t_{tot} = total travel time of surface water and groundwater,

t_s = surface-lateral travel time within the area with surface runoff towards aquifer recharge zone,

t_v = vertical travel time from the ground surface to the saturated zone,

t_h = horizontal travel time to the source.

3 Study areas

The Beli Timok groundwater source supplies drinking water to the town of Zaječar in eastern Serbia. The source comprises 10 shallow wells, which draw on the alluvial deposits of the Beli Timok River (Figure 4.2a). The average thickness of this sand-and-gravel aquifer is about 3 m and the average hydraulic conductivity $K_f = 0.005$ m/s (Figure 4.2b). The sand and gravel deposits are overlain by clayey sediments of low permeability (Milojević *et al.*, 1976). The groundwater is hydraulically connected with surface water, and the river stage directly influences the groundwater level and the production regime of the groundwater source. The thickness of the unsaturated zone is generally 2.5–3.5 m, or up to 5 m during the summer period of high water demand. The average production capacity of the source is about 30 l/s.

The Fišerov Salaš groundwater source is located in northwestern Serbia and supplies the city of Ruma. The average capacity of this source is 72 l/s. Groundwater is extracted from eight production wells (Figure 4.3a) whose depth ranges from 141 to 151 m below ground

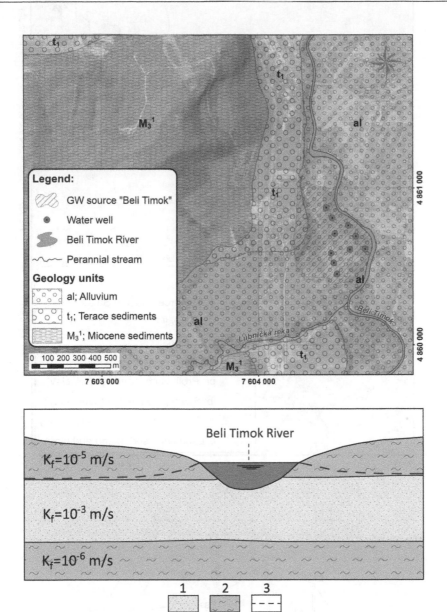

Figure 4.2 Geological map of Beli Timok groundwater source (a) and hydrogeological cross-section
(b) 1. Sand and gravel (saturated zone); 2. Low-permeable rocks; 3. Groundwater level.

surface (bgs). Well screens are installed at 110–115 m in a sandy layer (Figure 4.3b). The
hydraulic conductivity of the fine-grained sands is about 0.0003–0.0024 m/s (Josipović and
Soro, 2012). Initially, when the first wells were installed, the piezometric head was 4 m bgs,
but it is currently about 30 m bgs. Exploratory activities have revealed that the piezometric
head is mostly influenced by the groundwater abstraction regime.

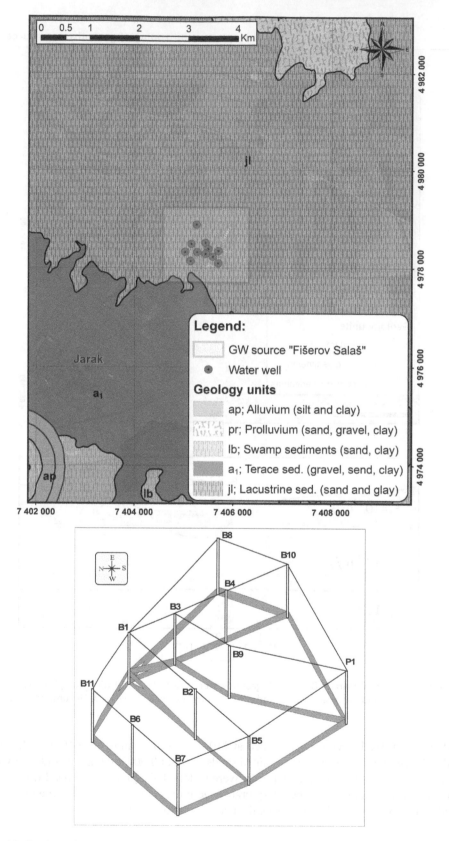

Figure 4.3 Geological map of Fišerov Salaš groundwater source (a), and fence diagram of the sandy aquifer and positions of wells (b) (Pušić, 1990).

4 Results

The conceptual model for vulnerability assessment of the Beli Timok source was based on known recharge-discharge relations. Three travel time components were analysed: time of horizontal groundwater travel through alluvial sediments to the water supply wells t_h, time of vertical groundwater travel through the overlaying strata t_v, and surface water travel time from a nearby hill to the alluvial plain t_s (Figure 4.4).

Horizontal groundwater flow was assessed using the ArcGIS Groundwater Toolbox (Esri, 2014). Several maps had previously been prepared, depicting the thickness of the saturated zone, transmissivity, effective porosity and groundwater level. The Darcy Flow tool (Esri, 2014) was used to create Flow Direction and Flow Magnitude maps. The Particle Track tool was then applied to calculate the horizontal travel times of groundwater to the production wells. The resulting pathways (flow lines) were used to generate a horizontal travel-time map (Figure 4.5).

The vertical travel-time components (t_v), from the ground surface to the saturated zone, were estimated for each specific point (pixel) in the area covered by the alluvial aquifer. The *Time-Input method* (Kralik and Keimel, 2003) was used to calculate the travel time through each layer in the unsaturated zone (the Time factor). The Input factor, which represents the recharge rate, was also estimated, by considering the amount of precipitation, solar radiation, vegetation, terrain slope, soil type, etc. Overlying of these two factors provides the vertical travel time (t_v) map shown in Figure 4.6.

The surface-lateral travel times (ts) were also calculated at each point in the area of surface runoff to the alluvial aquifer. The surface-lateral travel time is the sum of the sheet flow, shallow concentrated flow and channel flow components. These travel times can be calculated using Manning's equations for surface water flow. The necessary layers (length, terrain slope and roughness, and precipitation intensity) were prepared and analysed using ArcGIS Spatial Analyst – Hydrology Tools (Esri, 2014). The surface water travel-time map was generated using the Flow Length tool in combination with inverse values of surface water velocity (Figure 4.7).

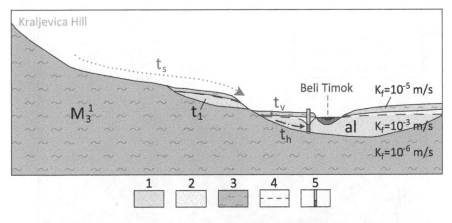

Figure 4.4 Conceptual TDM for vulnerability assessment of Beli Timok groundwater source I. Sand and gravel (saturated zone); 2. Sand and gravel (unsaturated zone); 3. Low-permeability rocks; 4. Groundwater level; 5. Water supply well.

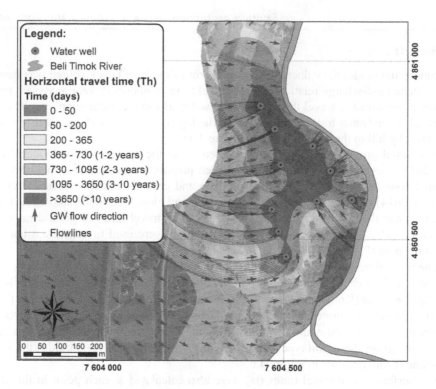

Figure 4.5 Horizontal travel time (t$_h$) through the saturated zone.

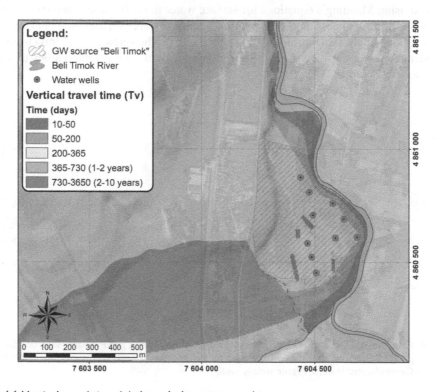

Figure 4.6 Vertical travel time (t$_v$) through the unsaturated zone.

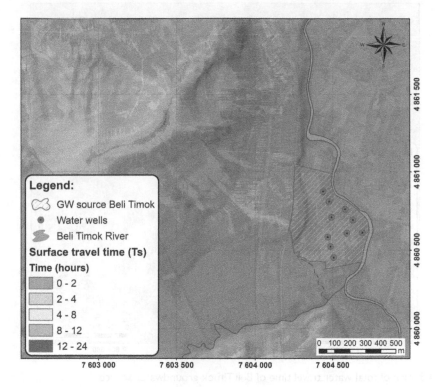

Figure 4.7 Map of surface-lateral travel times (t$_s$).

The actual groundwater source vulnerability map (Figure 4.8) was produced by overlying the maps that represented the three water travel time components and applying Equation 1. The resulting map represents the total water travel time from the ground surface to the water supply wells. The calculated travel times in the area close to the groundwater source were between 200 days and 2 years, due to the travel time through the unsaturated zone. Short travel times were obtained in the area close to the riverbank and abandoned artificial recharge basins, with a relatively thin unsaturated zone.

The conceptual model for vulnerability assessment of the deep confined aquifer at Fišerov Salaš was based on existing hydrogeologic units and the piezometric heads. The main challenge was to define the protection role of the overlying stratum, as impacted by groundwater levels under extraction conditions. Some vulnerability assessment methods like COP, PI and GOD (Vias *et al.*, 2006; Goldscheider *et al.*, 2000; Foster *et al.*, 2002) take into account the existence of confined aquifers, but none evaluate the actual piezometric head conditions, including the captured zone.

The conceptual model (Figure 4.9) considered two vertical travel-time components, whose directions were opposite. The first component (t$_{v1}$) represented the time needed for groundwater to travel downward through all the overlying strata to the aquifer (disregarding piezometric pressure). The second travel time (t$_{v2}$) was the time needed for groundwater to travel upward from the aquifer to the piezometric surface (or ground surface if the piezometric head is higher than the elevation of the ground surface). The vertical travel time (t$_v$) was

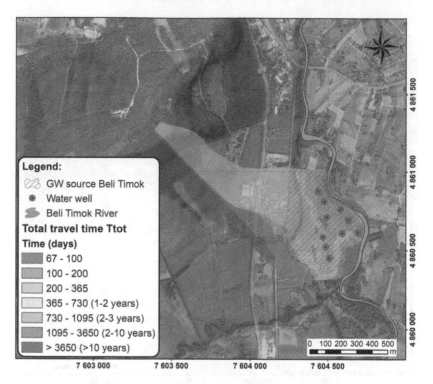

Figure 4.8 Map of total water travel time of Beli Timok groundwater source.

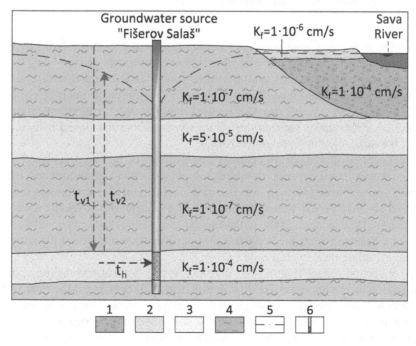

Figure 4.9 Conceptual TDM for vulnerability assessment of "Fišerov Salaš" groundwater source 1. Sand and gravel (alluvial aquifer); 2. Sand and gravel (unsaturated zone); 3. Low-permeable rocks; 4. Groundwater/piezometric level; 5. Water supply well.

calculated as the sum of the two components ($t_{v1}+t_{v1}$). This approach takes full account of the protective function of all overlying strata, as well as the existence of piezometric heads (particularly in the area of the groundwater source where the heads are drawn down).

Horizontal groundwater flow was assessed using a numerical model created in Visual Modflow. Groundwater flow was simulated at a source discharge capacity of 72 l/s. The flow paths around the wells were generated at different time steps. As a result, a map of travel time classes from 50 days to 5 years was produced (Figure 4.10a).

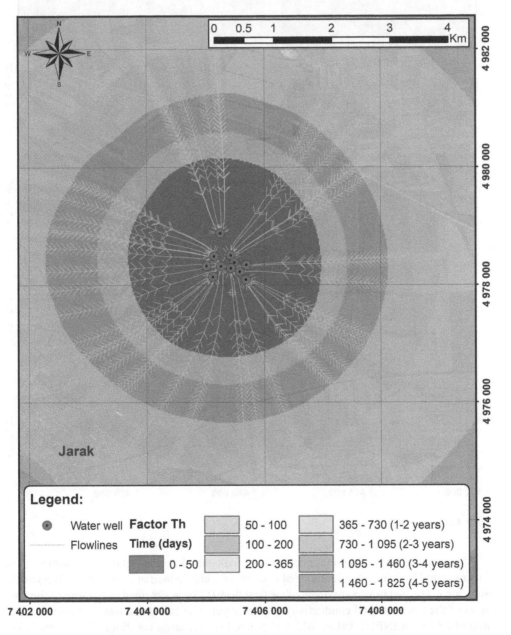

Figure 4.10 Maps of horizontal travel time – t_h (a) and vertical travel time – t_v (b).

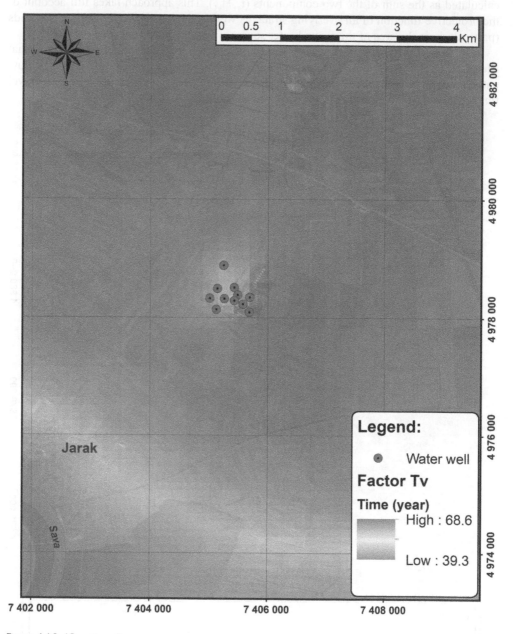

Figure 4.10 (Continued)

The vertical travel time components were calculated in an ArcGIS environment. The thickness and hydraulic conductivity of each layer were defined at each point (pixel) of the study area. The first time component of vertical flow (t_{v1}) was obtained as the sum of the quotients of the thickness and conductivity of each layer. The calculated travel times were then multiplied by an INPUT value, which depended on recharge conditions in the area. The second time component of vertical flow (t_{v2}) was calculated for those layers which spread between the aquifer and the piezometric head at a given point (pixel). The piezometric heads used in the calculations were a result of normal groundwater abstraction from the groundwater source. The map of vertical travel time (Figure 4.10b) was obtained by summing these two travel time components ($t_{v1}+t_{v1}$). The resulting map showed the time of vertical flow to be in the range from 40 to 70 years, greatest in the northern part of the source, where the thickness of overlying strata increases.

The source vulnerability map was created by overlaying the times for vertical and horizontal flow in each point/pixel of the study area (Figure 4.11). The surface-water travel time was disregarded since the recharge area is far from the source and the surface-water travel

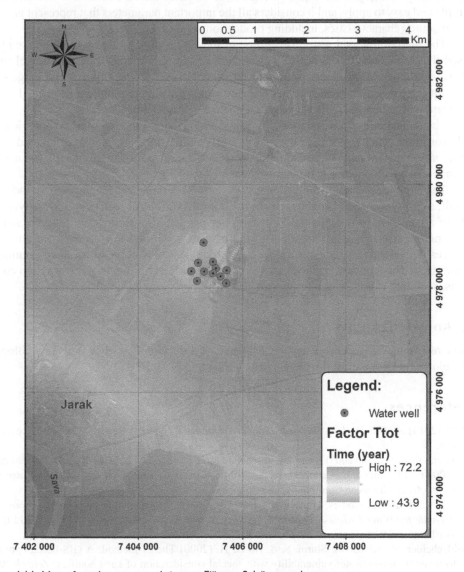

Figure 4.11 Map of total water travel time at Fišerov Salaš groundwater source.

times were insignificant, compared to the other time components. The map of the total travel time was very similar to the map of the vertical travel time, due to the dominant protection role of the overlaying strata. It is important to note that the groundwater abstraction regime and piezometric head variations neither had a significant effect on the total travel times nor the groundwater vulnerability.

5   Conclusion

The application of the TDM for the groundwater vulnerability assessment of two groundwater sources in unconsolidated sediments demonstrates that this method can be applied successfully to intergranular aquifers, regardless of the confining conditions. The method is simple and easy to apply, and it considers all the important parameters that represent groundwater source characteristics, including production conditions.

The map of the total water travel time of the Beli Timok groundwater source emphasised the importance of the protective function of the unsaturated zone. The travel times of water from the ground surface to the water supply wells are significantly prolonged and consequently the sanitary protection zones are smaller, compared to those suggested by the standard or commonly applied approach for sanitary zone delineation.

The vulnerability map of the Fišerov Salaš groundwater source showed favourable natural protection conditions due to the fact that the aquifer is confined. The calculated travel times indicated that the middle and outer protection zones could be significantly reduced, or even left out, with only basic protection measures in place.

The groundwater vulnerability maps are based on travel time components and, as a result, sanitary protection zoning can readily comply with local legislation. Moreover, one of the advantages of this method is that it does not require detailed and time-consuming exploration, which is very important in the case of small and low-capacity groundwater sources. Travel time calculations reduce subjectivity since the usage of parametric ranges is minimized. The proposed method can be applied to different groundwater extraction conditions, such that various groundwater protection scenarios can be simulated.

Acknowledgments

This research was supported by the Ministry of Education, Science and Technological Development (as a part of the Projects No. 43004 and No. 176022).

References

ESRI (2014) *ArcGIS Desktop: Release 10.2*. Environmental Systems Research Institute, Redlands, CA.

European Commission (2007) Common implementation strategy for the water framework directive: (2000/60/EC). In: *Guidance Document No. 16: Guidance on Groundwater in Drinking Water Protected Areas*. European Community, Luxembourg.

Foster, S., Hirata, R., Gomes, D., D'Elia, M. & Paris, M. (2002) *Groundwater Quality Protection: A Guide for Water Utilities, Municipal Authorities, and Environment Agencies*. The World Bank, Washington, DC.

Goldscheider, N., Klute, M., Sturm, S. & Hötzl, H. (2000) The PI method: A GIS-based approach to mapping groundwater vulnerability with special consideration of karst aquifers. *Zeitschrift für Angewandte Geologie*, 46(3), 157–166.

Josipović, J. & Soro, A. (2012) *Groundwater in Vojvodina.* Jaroslav Černi Water Institute, Belgrade (In Serbian).

Kralik, M. & Keimel, T. (2003) Time-input, an innovative groundwater-vulnerability assessment scheme: Application to an alpine test site. *Environmental Geology,* 44, 679–686.

Kresic, N. (2007) *Hydrology and Groundwater Modelling,* 2nd edition. (807p). CRC Press, Taylor & Francis Group, Boca Raton, FL.

Milojević, N. *et al.* (1976) *Geology of Serbia,* Book. VIII-1, Hydrogeology. Zavod za regionalnu geologiju i paleontologiju, Belgrade (In Serbian).

Parise, M., Ravbar, N., Živanović, V., Mikszewski, A., Kresic, N., Mádl-Szőnyi, J. & Kukurić, N. (2015) Hazards in Karst and managing water resources quality. In: Stevanovic, Z. (ed.) *Karst Aquifers: Characterization and Engineering, Professional Practice in Earth Sciences.* Springer International Publishing, Switzerland.

Pušić, M. (1990) *Artificial Recharge Od Deep Intergranular Aquifers.* PhD Thesis. Faculty of Mining and Geology, Belgrade (In Serbian).

Van Waegeningh, H.G. (1985) Overview of protection of groundwater quality. In: Matthess, G., Foster, S.S.D. & Skinner, A.C. (eds.) *Theoretical Background, Hydrogeology and Practice of Groundwater Protection Zones.* International Contributions to Hydrogeology, IAH, Volume 6. Verlag Heinz Heise, Hannover.

Vias, J.M., Andreo, B., Perles, M.J., Carrasco, F., Vadillo, I. & Jimenez, P. (2006) Proposed method for groundwater vulnerability mapping in carbonate (karstic) aquifers: The COP method. *Hydrogeology Journal,* 2006–14, 912–925.

Živanović, V., Jemcov, I., Dragišić, V., Atanacković, N. & Magazinović, S. (2016) Karst groundwater source protection based on the time-dependent vulnerability assessment model: Crnica springs case study, Eastern Serbia. *Environmental Earth Science,* 75, 1224. doi:10.1007/s12665-016-6018-2.

Zwahlen, F. (ed.) (2004) Vulnerability and Risk Mapping for the Protection of Carbonate (Karstic) Aquifers: Final Report COST Action 620. European Commission, Directorate-General for Research, Brüssel, Luxemburg.

Factors affecting vulnerability assessment – from scientific concept to practical application

Chapter 5

Meander effect on river-aquifer interactions

U. Boyraz & C.M. Kazezyılmaz-Alhan

I Introduction

Groundwater is a vital component of catchment hydrology. It contributes to the water supply significantly from wells and galleries. Both groundwater quality and quantity are affected by surface water due to the interactions between surface and groundwater. Therefore, sustainable management of the groundwater resources is crucial. For this purpose, the hydrodynamics of groundwater should be determined in detail (Winter *et al.*, 1998; Kania *et al.*, 2006). Surface water/groundwater interactions are significant and need to be taken into account in determining the hydrodynamic processes of groundwater. The surface water/groundwater interaction may be defined as a flow between surface water and groundwater and its characteristics depend upon the topographic features, hydraulic head difference and any other hydrodynamic conditions (Winter, 1999; Packman *et al.*, 2004; Kazezyılmaz-Alhan *et al.*, 2007). Contaminant transport occurs between the two water bodies with the interactions. As a result, the interaction between aquifer and surface water such as streams, lakes and wetlands, affects the hydrological behaviour and the contaminant transport mechanisms of the system (Kazezyılmaz-Alhan and Medina, 2008; Gooseff, 2010; Boano *et al.*, 2014).

Stream-aquifer systems are one of the most common formations where surface water/groundwater interaction is observed. Many field and laboratory studies show the existence of flow between stream and aquifer (Harvey and Bencala, 1993; Tonina and Buffington, 2009). Researchers have studied stream-aquifer interactions considering different aspects but there are still unresolved issues, which need to be investigated with further research studies. Stream shape plays an important role in identifying the hydrodynamics of interactions (Cardenas, 2009; Hester and Doyle, 2008). Boyraz and Kazezyılmaz-Alhan (2014) showed the significance of stream shape on interactions by determining the hyporheic flow rate for simple straight stream, streams with tributaries and meandering streams. According to their results, meandering streams have fluctuations in interaction rates which mean that the flow direction changes frequently. Boano *et al.* (2006) discuss the flow which is observed in a stream of sinusoidal shape.

In this study, the hyporheic exchanges in a meandering stream-aquifer region are investigated. The meanders of the stream are defined by a sinusoidal function with different amplitudes which represent wide or narrow meanders. Meandering Stream Finite Difference (MSF) numerical model is developed to simulate the groundwater flow in the aquifer and determine the hyporheic flux between the stream and the aquifer. The model is applied to synthetic examples and to the Duke restored wetland site in North Carolina. The effects of meander sharpness on the groundwater head distribution and on the hyporheic flux are discussed.

2   Numerical modelling

There are many analytical and numerical studies on groundwater flow (Spanoudaki *et al.*, 2010). A portion of these studies focus on stream-aquifer interactions (Hunt, 1990; Hantush *et al.*, 2002; Hantush, 2005; Lal, 2001). In most of these studies, stream head is commonly used as a boundary condition to solve the groundwater flow equation. This approach considers surface water and groundwater as a combined system. The solutions differ according to the different type of stream head boundary conditions such as sinusoidal waves or constant head, and aquifer properties such as isotropic, homogeneous, anisotropic and heterogeneous. The general form of groundwater flow equation is given as follows (McDonald and Harbaugh, 1988):

$$\frac{\partial}{\partial x}\left(K_{xx}\frac{\partial h}{\partial x}\right)+\frac{\partial}{\partial y}\left(K_{yy}\frac{\partial h}{\partial y}\right)+\frac{\partial}{\partial z}\left(K_{zz}\frac{\partial h}{\partial z}\right)+W=S_{s}\frac{\partial h}{\partial t} \tag{5.1}$$

where K_{xx}, K_{yy} and K_{zz} are hydraulic conductivities in x,y, and z directions (L/T), h is the hydraulic head (L), W is the source/sink term ($1/T$) ($W>0$ means inflow, $W<0$ means outflow), S_{s} is the specific storage ($1/L$), and t is time (T). To calculate the hyporheic flow rate, Darcy's Law is applied between stream head and groundwater head as follows (Prudic, 1989):

$$Q_{int\,eraction}=-K_{bed}\frac{\partial h}{\partial z}\times A \tag{5.2}$$

where Q is the hyporheic flow rate (L^3/T), K_{bed} is the streambed hydraulic conductivity (L/T), and A is the cross-section (L^2).

2.1   Conceptual model

In this study, a meandering stream-aquifer system is considered. The model is generated for a 2-D, homogenous and isotropic aquifer. The general form of groundwater equation reduces to the following form with these assumptions:

$$\left(\frac{\partial^{2}h}{\partial x^{2}}+\frac{\partial^{2}h}{\partial y^{2}}\right)=\frac{S_{s}}{K}\frac{\partial h}{\partial t} \tag{5.3}$$

Stream is defined as the boundary condition and the rest of the boundaries are defined as no-flux boundaries, i.e. Neumann boundary conditions.

Figure 5.1 shows the considered problem domain and its boundaries. Stream coordinates are calculated by using a sinusoidal function given as follows:

$$x=\Omega-\Omega\times\sin\left(\frac{2\pi y}{L_{y}}\right) \tag{5.4}$$

where Ω is the radius of the curvature. Equation 5.4 is used to generate a meander shape as shown in the figure which is placed along the y-direction of the aquifer. For different

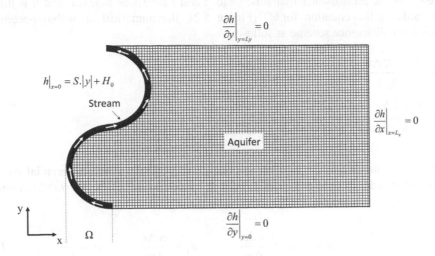

Figure 5.1 Meandering stream-aquifer model and the boundary conditions.

curvature radius and aquifer length, different meandering stream-aquifer models can be obtained by using this equation. As seen in the figure, hydraulic head in the stream is defined as linearly sloped.

2.2   Meandering Stream Finite Difference (MSF) model

A numerical model called as "Meandering Stream Finite Difference" (MSF) is developed by using finite difference scheme. The numerical algorithm is capable of solving two dimensional groundwater flow equations under transient conditions. This model is used to solve the set of Equations which represent a meandering stream-aquifer system. Discretisation of the differentials is shown in Equations 5.5–5.8 and the discretised form of groundwater flow equation is given in Equation 5.9.

$$\left.\frac{\partial h}{\partial t}\right|_{i} = \frac{h_{i,j}^{n+1} - h_{i,j}^{n}}{\Delta t} \tag{5.5}$$

$$\left.\frac{\partial^{2} h}{\partial x^{2}}\right|_{t}^{n} = \frac{h_{i+1,j}^{n} - 2h_{i,j}^{n} + h_{i-1,j}^{n}}{\Delta x^{2}} \tag{5.6}$$

$$\left.\frac{\partial^{2} h}{\partial y^{2}}\right|_{i,j}^{n} = \frac{h_{i,j+1}^{n} - 2h_{i,j}^{n} + h_{i,j-1}^{n}}{\Delta y^{2}} \tag{5.7}$$

$$K\left(\frac{h_{i+1,j}^{n} - 2h_{i,j}^{n} + h_{i-1,j}^{n}}{\Delta x^{2}} + \frac{h_{i,j+1}^{n} - 2h_{i,j}^{n} + h_{i-1,j}^{n}}{\Delta y^{2}}\right) = S_{s}\left(\frac{h_{i,j}^{n+1} - h_{i,j}^{n}}{\Delta t}\right) \tag{5.8}$$

where $h_{i,j}^{n+1}$ is the groundwater hydraulic head, i and j are space indexes, and n is the time index. Solving the equation for $h_{i,j}^{n+1}$ (Figure 5.2), the numerical algorithm becomes an explicit finite difference scheme as follows:

$$h_{i,j}^{n+1} = \frac{K\Delta t}{S_S \Delta x^2}\left(h_{i+1,j}^{n} + h_{i-1,j}^{n+1}\right) + \frac{K\Delta t}{S_S \Delta y^2}\left(h_{i,j+1}^{n} + h_{i,j-1}^{n}\right) - \cdots$$
$$\left(\frac{2K\Delta t\left(\Delta x^2 + \Delta y^2\right)}{S_S \Delta x^2 \Delta y^2}\right)h_{i,j}^{n} + h_{i,j}^{n} \qquad (5.9)$$

An explicit finite difference solution is conditionally stable. For a differential equation, which is first order in time and second order in space (Δt, Δx^2, Δy^2), *the stability condition $(d_x + d_y)$ is defined as follows:*

$$(d_x + d_y) \leq \frac{1}{2} \quad where \quad d_x = \frac{\alpha\Delta t}{\Delta x^2} \qquad d_y = \frac{\alpha\Delta t}{\Delta x^2} \qquad \alpha = \frac{K}{S_S} \qquad (5.10)$$

MATLAB is used for MSF simulation. MSF uses boundary matrixes as inputs; calculates the groundwater head distribution, and stream-aquifer interaction rates in time as outputs. MSF is also capable of making comparison of the calculated data with the experimental data. It calculates the error between the numerical model and the experiment and makes calibration for storage parameters.

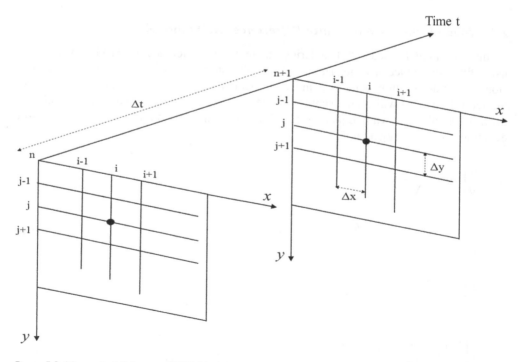

Figure 5.2 Numerical Scheme of MSF Model

2.3   *Visual MODFLOW (VMOD)*

VMOD calculates the groundwater flow in three dimensions by using finite difference numerical methods. A 3-D model of an aquifer, which is characterised by the aquifer parameters such as hydraulic conductivity, porosity and specific storage, can be generated and MODFLOW module solves the groundwater flow rate and head distribution under constant and variable head boundaries for both steady state and transient conditions. MODFLOW also uses stream water level as a boundary, and solves the interaction rate through the continuous saturated zone. In this study, VMOD is used only to verify the MSF model results. To create a 2-D model, aquifer should have only one layer in VMOD. The aquifer properties explained in the conceptual model are defined in VMOD to be able to compare VMOD results with MSF results. Meandering stream is generated by STR module of VMOD which is able to calculate the interaction flux. STR module is also capable of routing the flow through the streambed.

3   Application and results

3.1   *Synthetic examples*

Synthetic examples are solved by using MSF numerical model to determine the effects of the meander type of streams on groundwater head distribution and hyporheic flow. The verification of MSF model is also shown in equation 5.11. Meandering streams are placed into a homogeneous and isotropic aquifer, whose dimensions are 1600 m $\times$ 1600 m in x and y directions. Hydraulic conductivity and specific storage of the aquifer are assumed as 10^{-5} m/s and 10^{-5} 1/m, respectively.

Two types of meander are considered within the scope of this study. In the first example, a wide meander is considered with a large curvature and the stream is defined with the following sinusoidal equation:

$$x = 280 - 280 \times \sin\left(\frac{2\pi y}{1600}\right) \tag{5.11}$$

The physical meaning of this sinusoidal equation is that the meander is placed along y-axis with a length of 1600 m and the tip of the curvature reaches the point located 280 m far from the centre in x-direction of the aquifer. Meandering stream is defined as a boundary condition to the aquifer and the coordinates of the points forming the boundary are calculated by using the equation The hydraulic head at the boundary cells are defined as a linear variation to simulate sloped water surface in the meandering stream. The hydraulic head is defined as 17 m and 10.98 m at upstream point (280,0) and at downstream point (280,1600), respectively. These values correspond to a head boundary slope of 0.003 m/m. Groundwater head values are calculated for each grid to generate the boundary of the numerical model. A boundary matrix is given as an input into the MSF model. For transient analysis, an initial condition is determined as equal to the lowest hydraulic head of the stream.

++++The groundwater head contours are obtained by using both MSF and VMOD, at 5th, 10th, 20th, and 40th days. Results show that the system reaches the steady conditions at 40th day. The verification of MSF results is done by solving the same example using VMOD models under transient conditions. STR module is used to simulate the stream and WHS solver is used for the numerical solution in VMOD model. Slight differences between the solutions of MSF and VMOD are observed. These differences are attributed to the neglected source/sink term W in the MSF model whereas this term is taken into account in VMOD via STR module.

For a better understanding of the meander effect, simulation results of a simple (straight) stream should be considered. Boyraz and Kazezyılmaz-Alhan (2018) showed that the groundwater head contours are symmetrical under steady state conditions in an aquifer with a simple (straight) stream boundary. In addition, the interaction flow rate distribution along the simple stream was antisymmetric with respect to the midpoint of the stream. In this study, groundwater head contours obtained in an aquifer bounded by a meandering stream are not symmetrical unlike a simple stream case. The contours with constant intervals are dense at upstream and sparse at downstream which is an indication of different flow characteristics at upstream and downstream areas because of the meander. Dense contours represent high groundwater velocity values. At no-flux boundaries, groundwater flow is slower than in the other areas.

Figure 5.4 shows the hyporheic flux between stream and aquifer. At upstream point, the flux rate is 0.023 m/day/m^2 and it decreases to 0.005 m/day/m^2 downstream. At the 5th day, the groundwater flow occurs mostly from stream to aquifer. The aquifer recharges

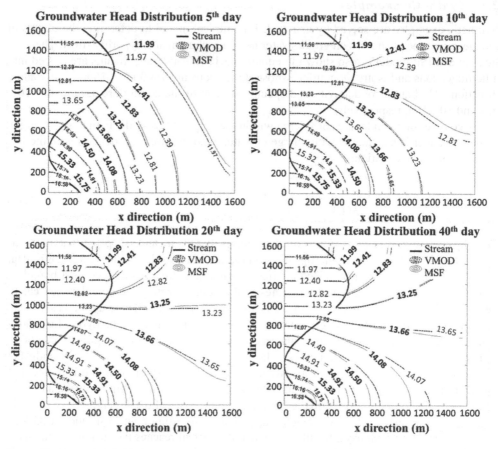

Figure 5.3 Groundwater head contours of a wide meander with a large curvature obtained by MSF and VMOD under transient conditions.

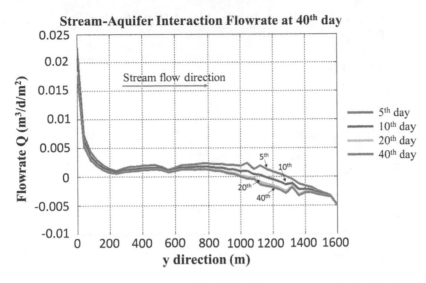

Figure 5.4 Interaction flow rate of a wide meander with a large curvature obtained by MSF.

along a major portion of the stream length. This recharge distance reduces while approaching the steady state condition. In steady state conditions, the aquifer recharges along 1000 m of stream length beginning at the upstream point. Then, the hyporheic flow reverses its direction from aquifer to stream for the last 600 m. In simple stream models, recharge/discharge distances were obtained as equal to each other in steady state conditions (Boyraz and Kazezyılmaz-Alhan, 2014). In meandering streams, the aquifer portion located in the inner part of the curvature is surrounded mostly by the stream and as a result, rapid head changes in short distances and complex flow paths are observed in this region. In simple straight streams, the aquifer is under the effect of the stream boundary only in a single direction which results in regular flow paths.

In the second example, the amplitude of the sinusoidal function is selected as 100 m which represents a narrow meander with a small curvature. In this case, the tip of the curvature reaches the point located 100 m far from the centre in x direction of the aquifer. The period of the function is 1600 m like the previous model. The same aquifer properties are used to investigate the curvature effects on hyporheic flow. The groundwater head contours obtained for the case of a narrow meander boundary are shown in equation 5.12 The results show that the groundwater head contours are more symmetrical than the results obtained for the case of a wide meander. This situation shows that the characteristics of a meandering stream lying in an aquifer decreases when curvature decreases. The groundwater flow velocity reaches the Neumann boundaries with less speed than the one in a wide meandering stream-aquifer system.

$$x = 100 - 100 \times \sin\left(\frac{2\pi y}{1600}\right) \tag{5.12}$$

Interaction fluxes are shown in Figure 5.6. Hyporheic flow rate is obtained as 8×10^{-4} m³/day/m² at upstream, and 4×10^{-4} m³/day/m² at downstream. The aquifer recharge is

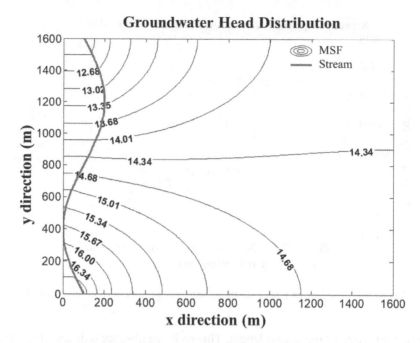

Figure 5.5 Groundwater head distribution of a narrow meander with a small curvature under steady state.

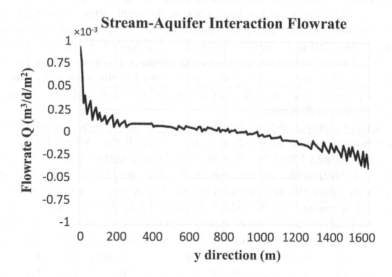

Figure 5.6 Interaction flow rate of a narrow meander with a small curvature obtained by MSF.

along the first 900 m of the stream. The aquifer discharge occurs along the last 700 m of the stream. As seen from these values, the big difference between the recharge/discharge distances diminishes for the case of a narrow meander. In case of a narrow meander, maximum interaction flow rate reduces almost 30 times in comparison with the case of a wide meander.

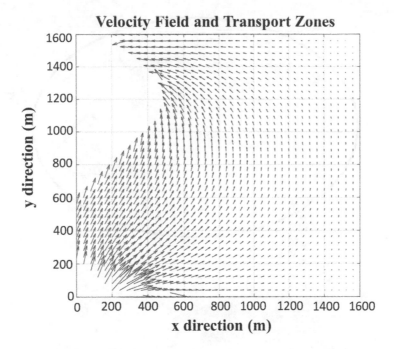

Figure 5.7 Velocity field and probable contaminant transport zones in meandering stream-aquifer systems.

These results strengthen the argument that as curvature of the meander decreases, the characteristics of the meandering stream lying on an aquifer also decrease.

Figure 5.7 shows the velocity field and flow paths of groundwater flow for meandering stream-aquifer systems under steady state conditions. Velocities of groundwater are very small towards the model boundaries. However, high velocity values are observed along the stream. The highest velocities are observed at the upstream point. Moreover, if a particle upstream passes to a transient zone, it travels through the aquifer within the curvature area and rejoins the stream downstream. Based on the velocity field of a meandering stream-aquifer system, it is concluded that a significant amount of contaminant transport between stream and groundwater is expected.

3.2   Study site application

The Duke restored wetland site and the restored stream in the wetland in Sandy Creek Watershed located in Duke Forest, North Carolina is selected as the study site to examine the capabilities of the MSF model. The study site including the surface topography (contours are plotted between 88.4 m and 100.6 m), the stream location, the well locations, the dam and the artificial lake is shown in Figure 5.8. Stream planform has two meanders at the upstream area, which were implemented in a former restoration project. The portion of the wetland shown with a circle in Figure 5.8 is modelled with MSF numerical model. Groundwater levels had been measured at 9 wells every other week. In addition, stream water depth had been measured regularly. The details of the wetland site and a wetland hydrology model

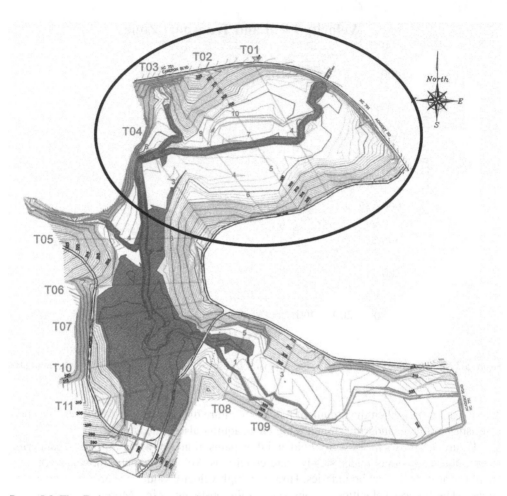

Figure 5.8 The Duke restored wetland site and the restored stream in the wetland in Sandy Creek Watershed (Kazezyılmaz-Alhan, 2005).

were presented by Kazezyılmaz-Alhan *et al.* (2007) and Kazezyılmaz-Alhan (2005). In this study, the stream-groundwater model is applied to the Duke Wetland Site and groundwater head contours are obtained. Then, the numerical results are compared with the collected data to see the performance of the MSF model.

In order to apply the MSF model to the study site, the area is discretised into 418 cells. First, the stream cells are determined and stream water level measurements are used as boundary condition in these cells. Neumann boundary condition is used at the side boundaries. The aquifer is assumed as homogeneous and isotropic. The streambed slope is 0.0038 and the hydraulic conductivity of the soil is $0.65 \times 10\text{--}3$ m/h. The steady state condition is considered. Stream level is 89.43 m at upstream and decreases linearly to 88.12 m at downstream. The groundwater head contours obtained by MSF model are shown in Figure 5.9.

In Figure 5.9, the line represents the stream and red boxes indicate the well locations. Grey cells show the inactive cells in the model. Groundwater head contours are close to each

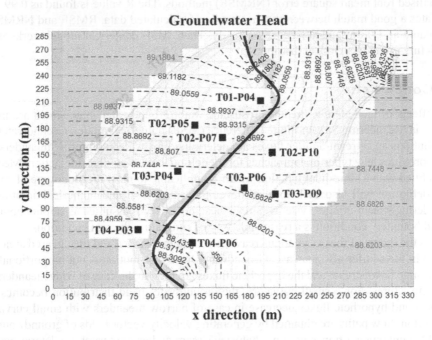

Figure 5.9 Groundwater head contours calculated by MSF model.

Table. 5.1 Measured and calculated groundwater level data.

Line	Well	Measured Data (m)	Calculated Data (m)	R^2	RMSE	NRMSE
T01	P04	89.87	89.06			
T02	P05	89.62	88.81			
T02	P07	89.61	88.87			
T02	P10	90.06	88.93			
T03	P04	88.80	88.68	0.99	1.06 m	0.012
T03	P06	89.72	88.68			
T03	P09	90.55	88.74			
T04	P03	89.40	88.45			
T04	P06	89.72	88.43			

other on the left side of the stream. They show a regular pattern around the straight segment of the stream and this regular pattern is broken around the meander segments. On the right side of the stream, the meander effect on the groundwater contours is more noticeable. The distance between the two head contours increases towards the right-hand boundary. Head contours are close to each other near the stream. They are close around the meander segments of the stream which represent fast groundwater flow and high interaction flow rates. The calculated data at the well locations is given in Table 5.1. The measurements at the wells and the error between the MSF model results and measured data are also presented in this table. The model efficiency is assessed with the R value, root mean square error (RMSE) and normalised root mean square error (NRMSE) methods. The R value is found as 0.99 which indicates a good match between the measured and calculated data. RMSE and NRMSE are calculated as 1.06 m and %1.12, respectively. Thus, MSF numerical model works well in predicting groundwater head distribution.

4   Conclusions

The effects of meandering streams on surface water/groundwater interactions are investigated via Meandering Stream Finite Difference (MSF) numerical model. Groundwater head distributions and stream-aquifer fluxes are determined for wide and narrow meanders. This study provides new types of approaches for modelling stream-aquifer interaction. Meanders are defined with a sinusoidal function which is applicable to most meander type of streams. This definition will also be helpful in developing analytical and numerical solutions for meandering stream-aquifer systems. Stream head boundary is assumed as sloping boundary; sloped boundary condition is a reasonable way of simulating the real conditions.

As the radius of the curvature increases, the impact distance and the covered area of the stream in the aquifer also increase. Groundwater head distribution is not proportional from upstream to downstream and the hyporheic fluxes are high in the case of wide meanders with large curvatures. For a small curvature, the groundwater head distribution becomes symmetrical and hyporheic fluxes decrease in case of narrow meanders with small curvatures. In addition, flow paths are obtained by generating velocity vector fields of groundwater, and possible contaminant transport zones between stream and groundwater are determined.

In order to show the applicability of MSF model to a field site, a portion of the stream-aquifer region at Duke restored wetland site is modelled and the calculated groundwater levels are compared with the measured data. It is concluded that MSF is capable of modelling groundwater head in meander types of stream-aquifer interaction.

This study provides an insight to the characteristics of meandering stream-aquifer systems. Different values of radius of curvature cause different groundwater flow behaviour in the aquifer. The results presented here will be helpful in site investigations and predictions on meandering stream-aquifer hydraulics. Classification of meanders depending upon the curvature types would be useful in future studies. Contaminant transport mechanism is a complex issue for meanders. Future study of this work may be on investigation of contaminant transport mechanisms in meandering stream-aquifer system including residential time and travel length of a contaminant.

5   Acknowledgement

The authors would like to express their gratitude to the anonymous reviewer. The authors would like to acknowledge also the Duke Wetland Center Case Studies Program and the NC Clean Water Management Trust Fund for funding water level data collections and, most importantly, funding the restoration of the stream/wetland lake complex in the Duke Forest.

References

Boano, F., Camporeale, C., Revelli, R. & Ridolfi, L. (2006) Sinuosity-driven hyporheic exchange in meandering rivers. *Geophysical Research Letters*, 33(18).
Boano, F., Harvey, J.W., Marion, A., Packman, A.I., Revelli, R., Ridolfi, L. & Wörman, A. (2014) Hyporheic flow and transport processes: Mechanisms, models, and biogeochemical implications. *Reviews of Geophysics*, 52. doi:10.1002/2012RG000417.

Boyraz, U. & Kazezyılmaz-Alhan, C.M. (2014) An investigation on the effect of geometric shape of streams on stream/ground water interactions and ground water flow. *Hydrology Research*, 45(4–5), 575–588. doi:10.2166/nh.2013.057.

Boyraz, U. & Kazezyılmaz-Alhan, C.M. (2018) Solutions for groundwater flow with sloping stream boundary: Analytical, numerical and experimental models. *Hydrology Research*, 49(4), 1120–1130.

Cardenas, M.B. (2009) Stream-aquifer interactions and hyporheic exchange in gaining and losing sinuous streams. *Water Resources Research*, 45, W06429. doi:10.1029/2008WR007651.

Gooseff, M.N. (2010) Defining hyporheic zones: Advancing our conceptual and operational definitions of where stream water and groundwater meet. *Geography Compass*, 4(8), 945–955.

Hantush, M.M. (2005) Modeling stream-aquifer interactions with linear response functions. *Journal of Hydrology*, 311(1–4), 59–79.

Hantush, M.M., Harada, M. & Marino, M.A. (2002) Hydraulics of stream flow routing with bank storage. *Journal of Hydrologic Engineering*, 7(1), 76–89.

Harvey, J.W. & Bencala, K.E. (1993) The effect of streambed topography on surface subsurface water exchange in mountain catchments. *Water Resources Research*, 29(1), 89–98. doi:10.1029/92WR01960.

Hester, E.T. & Doyle, M.W. (2008) In-stream geomorphic structures as drivers of hyporheic exchange. *Water Resources Research*, 44, W03417. doi:10.1029/2006WR005810.

Hunt, B. (1990) An approximation for the bank storage effect. *Water Resources Research*, 26(11), 2769–2775.

Kania, J., Haladus, A. & Witczak, S. (2006) On modelling of ground and surface water interactions. In: Baba, A., Howard, K.W.F. & Gunduz, O. (eds.) *Groundwater and Ecosystems*. Springer, Dordrecht.

Kazezyılmaz-Alhan, C.M. (2005) *A Wetland Model Incorporating Overland and Channel Flow, Solute Transport and Surface/Ground Water Interactions*. Ph.D. Thesis, Dept. of Civil and Env. Eng., Duke University, USA.

Kazezyılmaz-Alhan, C.M. & Medina, M.A., Jr. (2008) The effect of surface/ground water interactions on wetland sites with different characteristics. *Desalination*, 226, 298–305.

Kazezyılmaz-Alhan, C.M., Medina, M.A., Jr. & Richardson, C.J. (2007) A wetland hydrology and water quality model incorporating surface water/ground water interactions. *Water Resources Research*, 43(4), W04434.

Lal, A.M.W. (2001) Modification of canal flow due to stream-aquifer interaction. *Journal of Hydraulic Engineering*, 127(7), 567–576.

McDonald, M.G. & Harbaugh, A.W. (1988) *A Modular Three Dimensional Finite-Difference Groundwater Flow Model*. United States Government Printing Office, Washington, DC, USA, Report Number: 01-985-83961.

Packman, A.I., Salehin, M. & Mattia Zaramella, M. (2004) Hyporheic exchange with gravel beds: Basic hydrodynamic interactions and bedform-induced advective flows. *Journal of Hydraulic Engineering*, 130, 647–656.

Prudic, D.E. (1989) *Documentation of a Computer Program to Simulate Stream-Aquifer Relations Using a Modular, Finite-Difference, Ground-Water Flow Model*. U.S. Geological Survey, Open File Report. Report Number: 88-729. Available from: https://pubs.usgs.gov/of/1988/0729/report.pdf.

Spanoudaki, K., Nanou-Giannarou, A., Paschalinos, Y., Memos, C.D. & Stamou, A.I. (2010) Analytical solutions to the stream-aquifer interaction problem: A critical review. *Global Nest Journal*, 12(2), 126–139.

Tonina, D. & Buffington, J. (2009) Hyporheic exchange in mountain rivers. I: Mechanics and environmental effects. *Geography Compass*, 3. doi:10.1111/j.1749-8198.2009.00226.x.

Winter, T.C. (1999) Relation of streams, lakes, and wetlands to ground water flow systems. *Hydrogeology Journal*, 7, 28–45.

Winter, T.C., Harvey, J.W., Franke, O.L. & Alley, W.M. (1998) *Ground Water and Surface Water a Single Resource*. U.S. Geological Survey. Circular 1139. Available from: https://pubs.usgs.gov/circ/circ1139/.

Chapter 6

Comparison of pre-mining and post-mining conditions in an area impacted by coal mining as an aid to groundwater vulnerability assessment

K. David, W. Timms & R. Mitra

I Introduction

Quantitative assessment of groundwater intrinsic vulnerability is not possible without pre-mining data and is limited by lack of data on the characteristics and competence of the low permeability confining interbeds. Intrinsic vulnerability as defined by Stigter *et al.* (2006), relates to the impact of the change in groundwater levels/pressures and material properties on the value of the aquifer. The shallow aquifer is important as it provides baseflow to creeks and can influence the water balance of wetlands system, and hence ecosystem health. Shallow aquifers may also be important in the overall catchment water balance, with contributions to the Sydney Metropolitan water supply of immense value. Vulnerability of the shallow aquifer is particularly significant if strata deformation, depressurisation and dewatering have occurred below this unit and have caused changes in hydraulic properties resulting in changed pore pressures and ultimately changed groundwater flow conditions.

The role of massive conglomerate and sandstone units in surface subsidence in the Sydney Basin has been known for some time (McNally *et al.*, 1996), and is currently subject to further review (PSM, 2017; IEPMC, 2018). Massive rock units have been found to have strong bridging characteristics which reduce the impact on overburden behaviour and subsidence above the longwall panel (Whittaker and Reddish, 1989). These characteristics are important for the protection of the shallow aquifers from the direct effects of undermining. It is well known that fracturing that develops over the mined longwall panel results in changed hydraulic conductivity and groundwater flow (Mills, 2012) (Figure 6.1).

Strata Engineering (2003) found that the thickness of the overburden strata units was the most important controlling factor on maximum subsidence at the surface, compared with other longwall panel design parameters such as extraction height, panel width and orientation of panels. This is important as any deformation in the overburden will generate shearing and delamination into thinner units reducing their stiffness. Of particular importance for subsidence and groundwater behaviour is, therefore, the presence of thinly bedded siltstone, mudstone or claystone. The work by Ditton and Merrick (2014) offers an alternative method for subsurface fracture zone height prediction which takes into consideration geology in order to define the zone of disconnected fracturing, as discussed and critiqued recently by PSM (2017). In other words, identifying the extent of this zone allows further management options to be considered for the protection of overlying productive aquifers and surface water.

This paper focusses on the constrained zone, which is strata above the fractured zone (where significant disturbance and large downward groundwater flow occurs). Within this

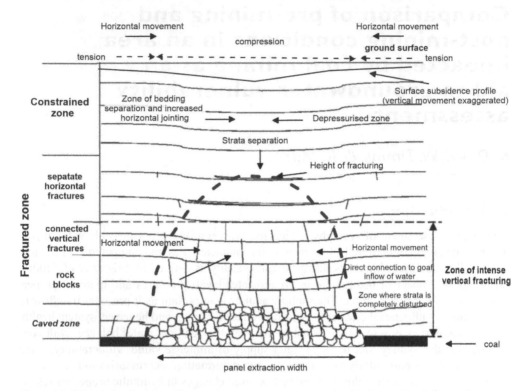

Figure 6.1 Schematic diagram of ground movement processes in the longwall mining environment (adapted from Holla and Barkley (2000))

constrained zone less ground movement occurs although stress propagation may extend to the surface (Figure 6.1) depending on the geometry of the panel, the thickness of the overburden and geology. Stress results in a strain which is associated with horizontal shearing and bedding separation resulting in redistribution of pressure and change to storage capacity. In order to study these phenomena, it is important to understand the change in porosity and *in situ* specific storage in the pre-mining environment to provide baseline and compare with post-mining conditions. This study, therefore, combines geophysical (bulk density) and groundwater methods (porosity and specific storage) to quantify changes in the constrained zone.

The ability to predict disturbance in the constrained zone on water-bearing strata could enable adaptive management that reduces the impact of mining on the overlaying groundwater systems. Understanding the change in hydraulic parameters in the constrained zone is critical, as this zone provides the protection of overlying aquifers, surface water and ecosystems in the surficial zone from excessive depressurisation and dewatering.

2   Background

The study area is located in the Southern Coalfields region in the Permian-Triassic age Sydney Basin, eastern Australia where several underground coal mines are operating. These mines operate in the proximity of the Sydney metropolitan water supply reservoirs and underneath the creeks and upland swamps. While aquifer vulnerability has not been assessed for this area, there are concerns for water losses and increased iron precipitation in surface water as a result of changed subsurface flow paths (McNally and Evans, 2007). Historical mining has resulted in dewatering of lower formations where connected fracturing exists, and depressurisation is occurring in the upper formations.

The Illawarra coal measures comprise of the Bulli seam, the main coal unit within the Permian sediments. Below the coal measures are sediments deposited in the marine environment while overlaying Triassic sediments are of fluvial origin. Tertiary igneous intrusions are present throughout the southern coalfields and occur as sills and dykes. Typical geology column is provided in Figure 6.2.

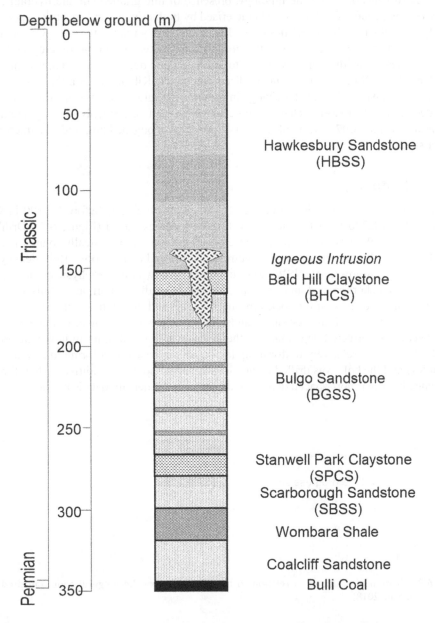

Figure 6.2 Stratigraphic column showing typical geology in the southern Sydney Basin

The Narrabeen group overlies the Illawarra Coal Measures and starts with Wombarra Shale/Claystone. The thickest unit in Narrabeen Group is the Bulgo Sandstone (BGSS) which is highly heterogeneous and comprises of a number of higher and lower permeability layers. The sandstone is interbedded with thin to medium thick layers of siltstone, mudstone and claystone with are generally of lower permeability than sandstone. The top of the group is represented by the regionally extensive Bald Hill claystone unit (BHCS). This unit is around 20 m thick and is considered an aquitard. The Hawkesbury sandstone (HBSS), which is the uppermost hydrostratigraphic unit, supports swamp ecosystems and water supply bores in the region and is characterised by good water quality. This unit is locally divided into three subunits based on the lithology, presence of fine-grained silt and hydraulic conductivity (Lee, 2000). The HBSS is characterised by both primary and secondary porosity.

Hydrogeological characteristics of the Basin are closely related to geology, topography and structural features. Recharge to the groundwater system occurs at outcrop and along open fractures, while discharge is related to low lying and deeply incised creeks and rivers and along the clifflines where strata are discontinued (McKibbin and Smith, 2000). Where coal extraction has not induced a change in groundwater gradients, upward flow can occur from deep strata. Increased recharge can occur where strata are steeply dipping and highly fractured (Pashin, 2007), with subsequent lateral flow along bedding and via fractures to deeper strata.

3   Methodology

Two drillholes were drilled in the centre of a longwall panel; one before (S2190 in 2012) and one after (S2220 in 2014) the longwall (LW9) was extracted (Figure 6.3). Drillholes S2190 and S2220 (at 50 m distance from each other) were geophysically logged immediately following drilling using density, neutron, gamma and caliper tools. Bulk density data (collected by Illawarra Coal, NSW) were obtained by downhole geophysics (gamma tool) which recorded short-spaced and long-spaced density readings. Formation bulk density is a function of the density of the rock-forming minerals and the fluid within the pore spaces and is a good porosity indicator since sandstone density is well known. Long-spaced density needs to be corrected depending on the groundwater and drilling fluid characteristics. The fluid density is relatively uniform (up to 500 µS/cm based on information from nearby boreholes) and the hole was drilled using clean water without any additives. Therefore, the difference between short-spaced density and compensated density data is minor.

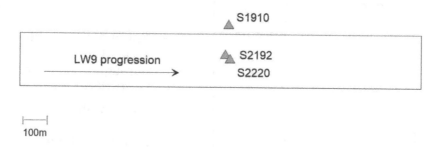

Figure 6.3 Location of drillholes in relation to the orientation of the longwall panel (adapted from David, 2018)

In 2008 a drillhole (S1910) was drilled at the northern edge of the LW9 and a string of eight vibrating wire piezometers (VWP) were installed at different depths in the borehole to measure the piezometric head at hourly intervals in the following strata: at surface and in the upper and lower HBSS, upper and lower BGSS, the underlying Scarborough Sandstone (SBSS) and at coal seam level. VWPs were fully grouted in, with electrical cables extending to surface to allow measurements. The piezometers provided hourly readings from August 2008 to October 2013. The reading was seriously impacted by mining after February 2013 when LW9 started. For this study, the upper six VWPs were used.

In order to estimate the influence of barometric pressure on pore pressure, selected time periods were used for the analysis. These periods were selected where data was not subject to other influences such as micro-seismic events resulting from coal extraction and fracturing. Hourly barometric pressure data was obtained from Australian Bureau of Meteorology (BoM) and site data. The corrected pore pressure (p) was calculated based on the following equation:

$$p = \left(p_o - B_{ave}\right) - LE\left(B - B_{ave}\right)$$
(6.1)

where p_o is the uncorrected absolute hydraulic head (kPa), B is the barometric pressure (kPa) for each data point, and B_{ave} is the average barometric pressure (kPa) over the entire time period. The calculated value of p is then plotted with time, along with the uncorrected hydraulic head data and barometric pressure, and the LE is adjusted until the graph p appears to be smooth and monotonically changing (Anochikwa et al., 2012; Smith et al., 2013). Particular attention was paid to data analysis for BHCS from the lower HBSS and upper BGSS. The influence of barometric pressure on water levels in bores has been observed in the past (Pascal, 1973; Rasmussen & Crawford, 1997; van der Kamp & Maathuis, 1991) and recently both barometric and earth tide influence were reviewed by McMillan et al. (2019). There is a significant difference between the magnitude observed in hydrographs of shallow water bores compared to deep confined aquifers. van der Kamp and Maathuis (1991) indicate that the pattern in deep confined aquifers does not reflect recharge through the confining layer but the changes in the total mechanical load on the groundwater system. In confined aquifers, the barometric pressure results in an instantaneous head response (Jacob, 1940) independent of lag as both confined aquifer and aquifer matrix take on the barometric pressure load and vertical flow is negligible (van der Kamp and Maathuis, 1991). This fluctuation in groundwater levels as a result of atmospheric pressure changes is termed loading efficiency and is defined as (van der Kamp and Gale, 1983):

$$LE = \frac{\alpha}{\alpha + n\beta}$$
(6.2)

Where α is vertical compressibility of aquifer skeleton (Pa^{-1}); β is water compressibility (Pa^{-1}), and n is the porosity. An increase in overburden pressure will result in a decrease in aquifer elasticity; therefore, barometric efficiency will be 100% where aquifer skeleton is not elastic. Loading efficiency (LE) (which is the response of pore pressures to external loading) can then be used to estimate the specific storage (S_s) by using the barometric efficiency (BE), porosity, water density (ρ) and gravity (g) as follows:

$$S_s = \frac{\rho g n \beta}{BE}$$
(6.3)

Where

$$BE = 1 - LE \tag{6.4}$$

Compressibility and specific storage (S_s) were computed (using equations 1 and 3) using hourly pore pressure data from S1910. The period of four months prior to the impact of mining (Sept to December 2008 -pre-mining) and after three panels were extracted (June 2012 to Jan 2013- mining) (Table 6.2) were used in this study.

The average of ratios method (Gonthier, 2007) was applied to selected time intervals where barometric pressure dependent pore pressure was devoid of any other influence (mainly pressure drop due to extraction). The average of ratios method calculated barometric efficiency, by dividing the difference in pore pressure with the difference in barometric pressure between two succeeding measurements (Gonthier, 2007).

Thus, loading efficiency in this paper was calculated using two methods: an average of ratios as described by Gonthier (2007) and secondly by visual adjustment (Smith *et al.*, 2013). The visual adjustment method developed by Smith *et al.* (2013) determines in situ compressibility and S_s by estimating loading efficiency of the strata from change in pore pressure due to barometric pressure fluctuations.

4   Results and interpretation

4.1   Density changes pre-mining and post-mining

A depth correlated dataset was used for the analysis of density changes. The first step involved correlation of compensated geophysical density log with post-mining porosity data measured directly on selected core samples (Figure 6.4a). Some smoothing and filtering of density dataset was undertaken as initial part of processing. Data is only presented and analysed from 40 m below ground level to the base of the hole because of the presence of bore casing above this depth. The short-spaced and long-spaced density logs show good agreement with changes in rock density when compared to visual inspection of core logs. Similarly, post-mining porosity and density results are inversely proportional, with an increase in porosity observed with a decrease in density (Figure 6.4a). Comparison of gamma and density data indicates the inversely proportional trend, with increase in gamma associated with a decrease in density (Figure 6.4b).

A combination of density and lithology correlation undertaken pre-mining (Figure 6.5) typically showed higher density in coarse-grained sandstone beds and lower density in fine-grained siltstone beds. A detailed review of pre- and post-mining density logs for the lower and upper BGSS (Figure 6.5) indicated that in the medium to coarse-grained materials, increased density was observed after longwall extraction. This density change was interpreted to be a result of mechanical impacts such as subsidence and induced strains. This change was consistent both in the strata close to the extracted panel, and also up to 160 m vertical distance above it (through the upper and lower BGSS).

However, in fine-grained siltstones, the change was reversed, with decreased density after longwall extraction. Similar to the change in coarse-grained sediments, this trend was noted both near the panel and further above it. The inferred increase in porosity associated with decreased density was attributed to the development of microfractures in fine-grained materials. Similar findings were reported by Stephenson *et al.* (1994), who described the

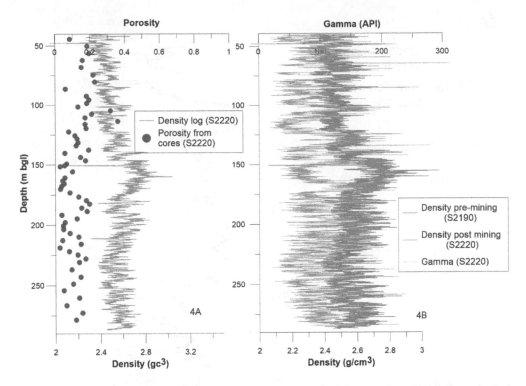

Figure 6.4 Depth correlation with post-mining porosity determined on core samples and geophysical compensated density log for S2220 post-mining 0–280m depth (4a) and pre-mining and post-mining density and gamma in the top 50m from surface (4b)

tightly-packed grain material expanding as a result of increasing strain, unlike poorly-sorted sediments where porosity and permeability decreased with increasing strain.

Although geophysical logs are measures of *in situ* properties, they are usually calibrated by regression (Paillet and Crowder, 1996). In this case, the regression analysis is undertaken with porosity data measured directly on the core samples (every 3 m along the borehole length) (Figure 6.4). The correlation between porosity measurements and density logs show general agreement, with a decrease in porosity associated with an increase in density.

Regression analysis (Figure 6.6) on the mean density and porosity dataset ($n = 64$, excluding surficial samples directly impacted by subsidence) showed scattering and a highly significant R^2-value of 0.16 ($P = 0.002$ at 95% confidence level). Within the confidence level of regression, the R^2-values varied from -0.005 to 0.306. The variability in density results is shown by error bars, based on 20 samples for each mean value (20 cm interval) (Figure 6.6). The confidence intervals were created for linear regression and show that the scattering was due to small-scale heterogeneity (cm size). The overall linear regression slope indicates a decrease in density with an increase in porosity, similar to that found when data was presented along the borehole length. It is also important to notice that the porosity measurement provides one value for a core sample about 20 cm in length, while density measurements are more frequent (every cm) and therefore capture changes in lithology and density along the sample's length.

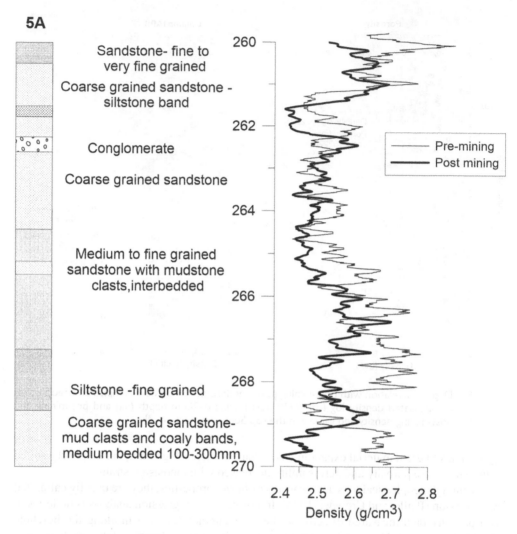

Figure 6.5 Depth-density correlation pre- and post-mining in the lower (5a) and upper (5b) BGSS, shown by the lithology log (David, 2018).

When geophysical and porosity (core) data were separated according to lithology and hydrostratigraphic unit, similar poor correlation ($R^2 = 0.15$) was obtained for HBSS and BHCS units where small numbers of samples were tested ($n = 5$). In addition to differences in the number of measurements per sample, a low R^2 may also be the result of microfracturing due to mechanical strain. Developed microfractures may be detected in density logs; however, these were not apparent in porosity data on cores due to sample preparation by crushing. This regression analysis shows that downhole geophysical density data could be more useful than the limited porosity data for identifying changes that occur in various lithology types due to stresses.

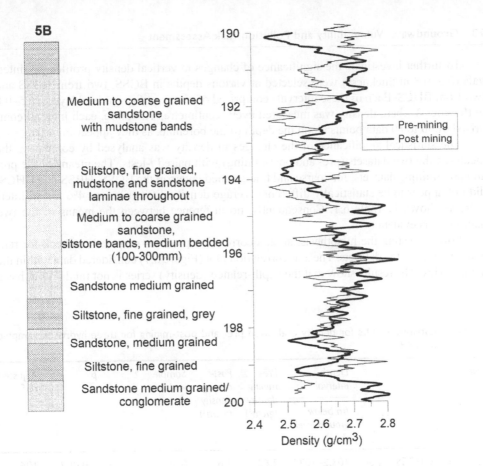

Figure 6.5 (Continued)

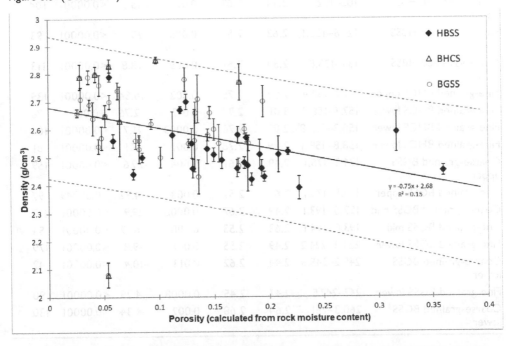

Figure 6.6 Regression of core porosity and density obtained from geophysical logs (S2220). Error bars represent the standard deviations for the 20 cm density intervals. The dotted line is the 95% confidence interval for linear regression. (David, 2018)

To further investigate the significance of changes in vertical density profiles, six intervals (0.5–0.8 m thickness) were selected at various depths in BGSS, two from HBSS and two from BHCS. Each of these intervals comprised of coarse or fine-grained layers (0.5–0.8 m thickness) where density was measured every centimetre. Therefore, each interval comprised of 50 to 80 data points along the depth of the borehole, as they appear in order.

The statistical significance of the changes in density was analysed by comparing the means of the two datasets (pre- and post-mining) with paired *t*-tests. The means of the pre- and post-mining datasets, for coarse and fine-grained layers in the BGSS, HBSS and BHCS, did not appear to be statistically different (average density difference was 4% for all intervals) as shown in Table 6.1. Consequently, no difference between the means of the two datasets was evident.

Prior to testing the hypothesis, an autocorrelation plot was prepared to check for randomness within the datasets. The autocorrelation plot (Figure 6.7) considered data within the depth series. The results show that the depth-related (density) series is not random but has a

Table 6.1 Statistical results for density evaluation pre- and post-mining for three hydrostratigraphic units +++

Unit	Depth interval (m below ground level (bgl))	Pre-mining density (g/cm³)	Post-mining density (g/cm³)	Sample variance	t	p	Adjusted dof
Fine-grained HBSS upper	103.2–103.9	2.62	2.6	0.001	−2.83	0.002	104
Coarse-grained HBSS upper	102–102.7	2.44	2.62	0.014	−15.2	<0.00001	104
Coarse-grained HBSS lower	121.6–122.1	2.63	2.5	0.005	25	<0.00001	93
Coarse-grained HBSS lower	135–135.8	2.39	2.51	0.005	−18.8	<0.00001	113
Fine-grained BHCS upper	151.7–152.4	2.82	2.75	0.002	16.54	<0.00001	135
Fine-grained BHCS lower	157.7–158.3	3.01	2.97	0.012	2.76	0.003	76
Fine-grained BHCS lower	158.3–158.8	2.85	2.76	0.0003	21.7	<0.00001	100
Fine-grained BHCS lower	158.8–159.3	2.89	2.78	0.001	17.1	<0.00001	81
Coarse-grained BGSS upper	174.3–175.1	2.49	2.61	0.004	−29.6	<0.00001	159
Fine-grained BGSS upper	175.5–176.1	2.61	2.57	0.003	4.92	0.00049	97
Coarse-grained BGSS mid	192.5–193.1	2.49	2.67	0.0008	−29.9	<0.00001	117
Fine-grained BGSS mid	193.6–194.1	2.63	2.53	0.008	6.12	<0.00001	54
Fine-grained BGSS lower	223.3–224.2	2.49	2.55	0.001	−9.8	<0.00001	179
Coarse-grained BGSS lower	245.2–245.6	2.44	2.62	0.013	−10.4	<0.00001	42
Fine-grained BGSS lower	247–247.5	2.47	2.45	0.0008	4.29	<0.00001	90
Coarse-grained BGSS lower	260.3–261	2.64	2.68	0.002	−4.34	<0.00001	130

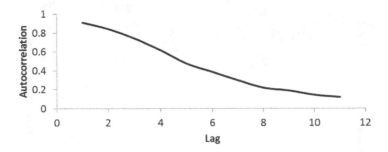

Figure 6.7 Autocorrelation plot showing the randomness of data for 11 lag periods (depth) (David, 2018)

high degree of autocorrelation between adjacent and near-adjacent observation points. This meant that the degrees of freedom had to be adjusted for unequal variances.

In order to test this hypothesis, a p-value had to be determined from the density dataset. The calculated p-value was significantly less than 0.05 for both coarse and fine-grained strata (Table 6.1) using the adjusted degrees of freedom (dof), indicating that the hypothesis was not valid and there was a significant difference between these two datasets.

Similar p-values were obtained from a comparison of the coarse and fine-grained layers from HBSS, at about 100 m and 120 m depths, and in BHCS (Table 6.1). Hence, on a small scale (cm up to one m) the changes in porosity are influenced by grain size and matrix characteristics, irrespective of the vertical distance from the mined panel.

To summarise, with upward strain propagation above the centre of the panel, the porosity of coarse and medium-grained material decreases while at the same time the porosity of the fine-grained material increases. This finding differs with previously reported (Booth, 2007; Kendorski, 1993) overall increase in porosity (matrix and fracture) in the post-mining regime. Based on the new findings in this paper, the actual net change in porosity in the overburden strata will be a more complex function including thickness of individual beds and their lithology, grain size and matrix characteristics.

4.2 Change in loading efficiency with mining

The results of the analysis show a pre-mining decrease in LE (and consequently S_s) with increasing depth, with the exception of SPCS whose LE is higher as it mainly comprises of claystone. This finding is as expected with increased overburden thickness. However, as the LW9 moves eastward and development approaches S1910, LE increases significantly and calculated S_s shows increases for BGSS and a slight increase for HBSS and SBCS. The implications of increased S_s over time is yet to be evaluated in groundwater models. The reason for variable S_s is attributed to straining and change in porosity from pre to post-mining. There is uncertainty associated with the evaluation of LE for strata below 247m depth due to low resolution in pore pressure readings (Figure 6.8d).

The average of ratio method provides higher loading efficiency in the pre-mining period, however post-mining it is significantly higher than visual method (Table 6.2). The reason for such discrepancy is due to an average of ratio method depending on the selection

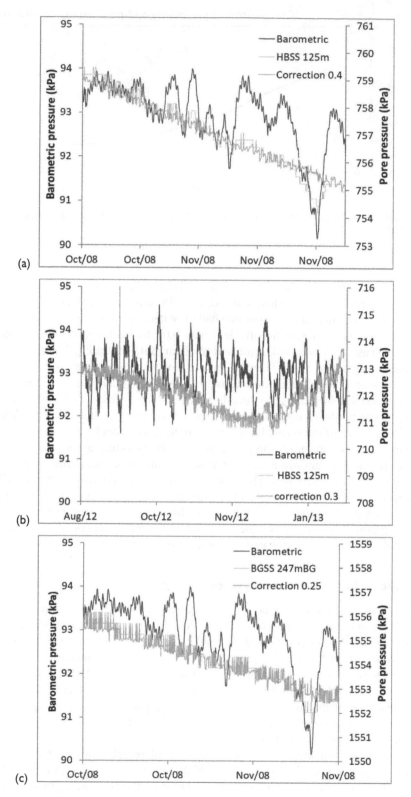

Figure 6.8 Loading efficiency correction for HBSS 125 m bgl (a and b) and BGSS 247 m bgl (c and d) pre-mining (a and c) and post-mining (b and d) (adapted from David *et al.*, 2017)

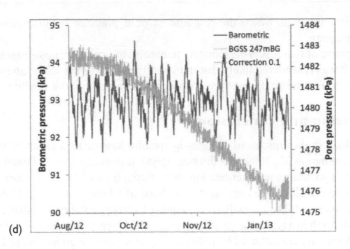

(d)

Figure 6.8 (Continued)

Table 6.2 Summary of compressibility and specific storage changes within hydrostratigraphic units estimated by barometric correction method and average of ratio method

	Hydrostratigraphic unit		HBSS	HBSS	BGSS	BGSS	SBCS
	Depth (m below ground level (bgl))		9.7	125	169	247	274
Pre-mining	n (effective)		0.11	0.1	0.03	0.07	0.13
	LE	Visual method (from corrected pore pressures)	0.67	0.64	0.4	0.35	0.32
	S_s		2.12E-06	1.57E-06	2.81E-07	5.90E-07	1.03E-06
	LE	Average of ratio method	0.6	0.5	0.4	0.5	0.5
	N		0.17	0.1	0.03	0.07	0.14
Post-mining	LE	Visual method (from corrected pore)	0.95	0.67	0.6	0.6	0.99
	S_s		1.53E-05	2.05E-06	5.64E-07	1.35E-06	6.31E-05
	LE	Average of ratio method	0.95	0.94	0.94	0.06	NA

of specific time intervals where the response in pore pressure is corresponding solely to barometric pressure.

The response of two SBCS piezometers to barometric pressure post-mining could not be determined. It was considered likely that the vicinity of the mined seam (362 m bgl) and the perturbations within the goaf zone were impacting the pore pressure readings.

5   Discussion and conclusion

Several previous studies discussed changes in specific storage as a result of subsidence and fracturing (Karaman *et al.*, 1999; Booth *et al*, 1998), however, there is limited understanding of compressibility changes in water-bearing strata through variable sedimentary sequence during mining. These findings are consistent with those of Chilingarian *et al.* (1995) who found that aquifer compressibility is smaller than aquitard compressibility and that compressibility decreases with depth in a typical sedimentary basin. Therefore, the fracturing and jointing resulting from undermining are followed by tight packing of sandstone grains and reduced porosity.

Below this zone, changes in density pre-mining and post-mining are not as uniform and an interesting trend was observed. On a detailed vertical scale (centimetre to meter), increased density was evident in coarse-grained material while in the fine-grained strata decreased density was observed. In contrast to previous reports of overall increase in porosity, these findings imply that the interlayering of coarse and fine-grained material may result in a change in hydraulic properties on a small scale with more complex influences on groundwater flow.

This paper estimated the change in loading efficiency and specific storage when significant deformation induced strains occur. The spatial and temporal changes in geomechanical strains within the constrained zone cause the redistribution of pore pressures, changes in compressibility and ultimately specific storage (David *et al.*, 2017).

In situ measurement of pore pressures and the impact of barometric pressure on these changes are likely to provide the most accurate data for the estimation of compressibility and specific storage. An overall increase in loading efficiency and compressibility was observed post-mining. This is a result of the location of the piezometer in the middle of the panel and vertical strains that occur relative to the panel.

Although subsidence may result in rapid loss of surface water above the longwall panel, a post-mining decrease in porosity in coarse-grained strata (immediately above the longwall panel) could be important in reducing the flow and leakage from upper to deeper hydrostratigraphic units.

In conclusion, determination of in situ specific storage, and any changes over time, is an important part of evaluating intrinsic aquifer vulnerability and could lead to improved groundwater models. The implications of increased specific storage over time is yet to be evaluated in groundwater models Together with detailed information on lithology, the significance of the extent of depressurisation (defined by specific storage) is a particularly important variable in areas where groundwater provides baseflow to creeks and supports wetland ecosystems and public water supplies.

Acknowledgement

The authors would like to thank BHPB (R. Walsh and the geology team) for providing the data for this paper and School of Minerals and Energy Resource Engineering, UNSW Australia for supporting the analysis of the rock cores. We are very thankful to K. Gamage,

M. Whelan and P. Chai for assisting with rock core testing and L. Barbour from the University of Saskatchewan for fruitful discussions in support of this paper.

References

Anochikwa, C.I., van der Kamp, G. & Barbour, L. (2012) Interpreting pore-water pressure changes induced by water table fluctuations and mechanical loading due to soil moisture changes. *Canadian Geotechnical Journal*, 49, 357–366.

Booth, C.J. (2007) Confined-unconfined changes above longwall coal mining due to increases in fracture porosity. *Environmental and Engineering Geoscience*, 13(4), 355–367.

Booth, C.J., Spande, E.D., Pattee, C.T., Miller, J.D. & Bertsch, L.P. (1998) Positive and negative impacts of longwall mine subsidence on a sandstone aquifer. *Environmental Geology*, 34(2/3).

Chilingarian, G.V., Rieke, H.H. & Donaldson, E.C. (1995) Compaction of argillaceous sediments. In: Chilingarian, G.V. (ed.) *Subsidence Due to Fluid Withdrawal, Dev. Pet. Sci.*, Volume 41, chap. 2. Elsevier Science, New York. pp. 47–164.

David, K. (2018) *Groundwater Changes in a Sedimentary Sequence Associated with Underground Mining: A Multidisciplinary Approach*. PhD thesis, UNSW Australia, Sydney, Australia.

David, K., Timms, W., Barbour, S. & Mitra, R. (2017) Tracking changes in the specific storage 1155 of overburden rock during longwall coal mining. *Journal of Hydrology*, 553, 304–320. doi:10. 1156 1016/j.jhydrol.2017.07.057.

Ditton, S. & Merrick, N. (2014) A new subsurface fracture height prediction model for longwall mines in the NSW coalfields. *Proceedings of Sydney Basin Symposium, Newcastle*.

Gonthier, G.J. (2007) *A Graphical Method for Estimation of Barometric Efficiency from Continuous Data-Concepts and Application to a Site in the Piedmont, Air Force Plant 6, Marietta.* (29p). US Geological Survey Scientific Investigation Report 2007-5111. Georgia. Available from: http://pubs. usgs.gov.sir/2007/5111/.

Holla, L. & Barkley, E. (2000) *Mine Subsidence in the Southern Coalfield*, NSW, Australia, Mineral Resources of NSW, Sydney.

Independent Expert Panel for Mining in the Catchment (IEPMC) (2018) Initial report on specific mining activities at the Metropolitan and Dendrobium coal mines. *Prepared for the NSW Department of Planning and Environment*. Available from: https://www.chiefscientist.nsw.gov.au/__data/assets/ pdf_file/0008/209357/IEPMC-Report_Term-of-Reference-1.pdf (Accessed 20 December 2018).

Jacob, C.E. (1940) On the flow of water in an elastic artesian aquifer. *Transactions American Geophysical Union*, 21, 574–586.

Karaman, A., Akhiev, S.S. & Carpenter, P.J. (1999) A new method of analysis of water-level response to a moving boundary of a longwall mine. *Water Resources Research*, 35(4), 1001–1010.

Kendorski, F.S. (1993) Effect of full-extraction mining on ground and surface waters. *Proceedings of the 12th International Conference on Ground Control in Mining, West Virginia University*, Morgantown, West Virginia, pp. 412–425.

Lee, J. (2000) Hydrogeology of the Hawkesbury Sandstone in the Southern Highlands on NSW in relation to Mesozoic horst-graben tectonics and stratigraphy. *Proceedings of the 34th Newcastle Symposium: Advances in the Study of the Sydney Basin*, Newcastle, NSW, Australia.

McKibbin, D. & Smith, P.C. (2000) *Sandstone Hydrogeology of the Sydney Region*, NSW Department of Land and Water Conservation, Sydney.

McMillan, T.C., Rau, G.C., Timms, W.A. & Andersen, M.S. (2019) Utilizing the impact of Earth and atmospheric tides on groundwater systems: A review reveals the future potential. *Reviews of Geophysics*, In press, May.

McNally, G.H. & Evans, R. (2007) *Impact of Longwall Mining on Surface Water and Groundwater, Southern Coalfield NSW*. Report prepared for NSW Department of Environment and Climate Change. eWater Cooperative Research Centre, Canberra.

McNally, G.H., Willey, P.L. & Creech, M. (1996) Geological factors influencing longwall-induced subsidence. *Symposium on Geology in Longwall Mining, November*.

Mills, K.W. (2012) Observations of ground movements within the overburden strata above long-wall panels and implications for groundwater impacts. *Proceedings of the 38th Symposium on the Advances in the Study of the Sydney Basin, Hunter Valley, May*.

Paillet, F.I. & Crowder, R.E. (1996) A generalised approach for the interpretation of geophysical well logs in ground-water studies-theory and application. *Groundwater*, 34(5).

Pascal, B. (1973) *The Physical Treatises of Pascal*. Octagon Books, New York.

Pashin, J.C. (2007) Hydrodynamics of coalbed methane reservoirs in the Black Warrior Basin: Key to understanding reservoir performance and environmental issues. *Applied Geochemistry*, 22, 2257–2272.

PSM (2017) *Height of Cracking: Dendrobium Area 3B, Dendrobium Mine*. Report for Department of Planning and Environment, PSM3021-002R, March. Available from: www.chiefscientist.nsw. gov.au/__data/assets/pdf_file/0008/209357/IEPMC-Report_Term-of-Reference-1.pdf (Accessed 2 February 2019).

Rasmussen, T.C. & Crawford, L.A. (1997) Identifying and removing barometric pressure effects in confined and unconfined aquifers, *Groundwater*, 35(3), May–June.

Smith, L.A., van der Kamp, G. & Hendry, J.M. (2013) A new technique for obtaining high resolution pore pressure records in thick claystone aquitards and its use to determine in situ compressibility. *Water Resources Research*, 49, 1–12. doi:10.1029/2012WR012166.

Stephenson, E.L., Maltman, A.J. & Knipe, R.J. (1994) Fluid flow in actively deforming sediments: 'Dynamic permeability' in accretionary prisms. In: Parnell, J. (ed.) *Geofluids: Origin, Migration and Evolution of Fluids in Sedimentary Basins*. Geological Society Special Publication No 78. pp. 113–125.

Stigter, T.Y., Ribeiro, L. & Carvalho Dill, A.M.M. (2006) Evaluation of an intrinsic and a specific vulnerability assessment method in comparison with groundwater salinisation and nitrate contamination levels in two agricultural regions in the south of Portugal. *Hydrogeology Journal*, 14, 79–99.

Strata Engineering (2003) *Review of Industry Subsidence Data in Relation to the Influence of Overburden Lithology on Subsidence and an Initial Assessment of a Sub-Surface Fracturing Model for Groundwater Analysis*. ACARP Report C10023, September.

van der Kamp, G. & Gale, J.E. (1983) Theory of earth tide and barometric effects in porous formations with compressible grains. *Water Resources Research*, 19, 538–544.

van der Kamp, G. & Maathuis, H. (1991) Annual fluctuations of groundwater levels as a result of loading by surface moisture. *Journal of Hydrology*, 127(1991), 137–152.

Whittaker, B.N. & Reddish, D.J. (1989) Subsidence overview, prediction and control. *Developments in Geotechnical Engineering*, 56, Elsevier.

Chapter 7

Nitrate migration in the regional groundwater recharge zone (Lwówek region, Poland)

K. Dragon

1 Introduction

The elucidation of groundwater flow systems can be assisted by the investigation of spatial and vertical changes in groundwater chemistry (Ophori and Toth, 1989; Hendry and Schwartz, 1990; Ochsenkuhn *et al.*, 1997; Coetsiers and Walraevens, 2006). Changes in groundwater chemistry can support understanding of groundwater flow systems especially in cases involving multiple aquifers, where the hydrogeology is more complex (Carillo-Rivera *et al.*, 1996; Al-Mashaikhi *et al.*, 2012; Dragon and Gorski, 2015). The most effective investigation is possible when the concentration of the chosen chemical tracer (for example nitrate) is characterised by clear spatial or vertical differentiation.

Groundwater contamination by nitrate is a significant environmental problem. There are increasing number of reports documenting nitrate pollution in many countries (e.g., Hudak, 2000; Rodvang and Simpkins, 2001; Chen *et al.*, 2005; Petitta *et al.*, 2009; Dragon *et al.*, 2016; Caschetto *et al.*, 2018). Nitrate pollution is also reported in the Polish literature (e.g., Górski *et al.*, 2019; Zurek *et al.*, 2010; Dragon, 2016; Czekaj *et al.*, 2016).

Groundwater pollution by nitrate is observed mainly in relatively shallow aquifers. However, in the regional recharge zones, characterised by distinct downward gradients caused by descending flow patterns, contamination can appear in deeper parts of the system (Chen *et al.*, 2005). This is especially true in unconfined regional aquifers, which are the most vulnerable to pollution from the surface. One factor that can activate or noticeably accelerate contaminant migration is groundwater extraction. This is because it can create a downward gradient, which enables a downward migration of contaminants (Dragon, 2013).

The main aim of this study was to investigate the behaviour of nitrate in relation to the groundwater flow system. Based on vertical changes of nitrate concentrations examined from multilevel observation wells, groundwater flow paths were investigated. The influence of groundwater exploitation on nitrates migration was also investigated.

2 Site description

The study area is located in an area of relatively high elevation in the Lwowek-Rakoniewice Rampart of western Poland (Figure 7.1).

For a detailed analysis of the water chemistry, the central part of the recharge area was selected, where there are conditions that cause the aquifer system to be the most vulnerable to groundwater contamination (Figure7.2). The lithology of the sediments is dominated by glacial and fluvioglacial deposits (Figure 7.2). The fluvioglacial sands and gravels form the aquifers, which create a multilayer aquifer system. The deepest aquifers

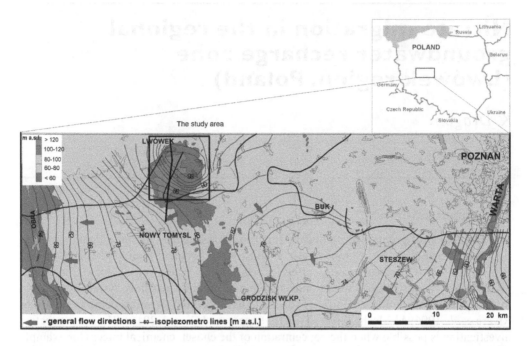

Figure 7.1 The study area (marked by the rectangle) on the isopiezometric contour map of and ground elevation. – line of the cross-section (Figure 7.2).

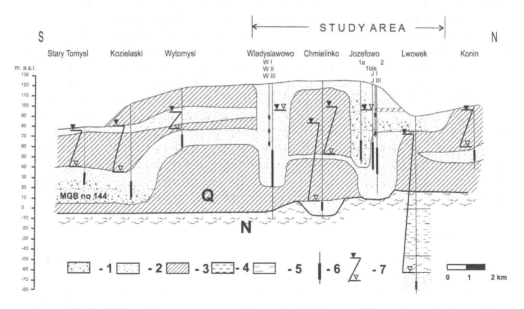

Figure 7.2 The hydrogeological cross-section (after Dragon, 2013, modified). 1 – Coarse sand and gravel, 2 – medium and fine sand, 3 – till, 4- clay, 5 – brown coal, 6- silt, 7 – the location of the well screen, 8- groundwater level, Q- Quaternary, N – Neogene

have a thickness of approximately 20 m and are composed of fine sands. The shallower aquifers have a variable thickness (between 5 and 20 m) and are composed primarily of fine and medium sands. The confining layers of these aquifers are glacial tills of variable thickness (Figure 7.2). In some regions (Jozefowo and Wladyslawowo) there is no aquitard and the aquifer is unconfined (Figure 7.2). Sands and gravels occur from the surface to a depth of more than 100 m. The unconfined parts of the aquifer system remain in hydraulic connection with the shallower confined aquifer system, whereas glacial tills cause the deepest aquifers to be isolated (Figure. 7.2).

It is shown on the isopiezometric contour map (Figure 7.1) that the selected region is located in the central part of the recharge area on the groundwater divide from which groundwater will either flow eastward (to the Warta river valley) or westward to the Obra river valley.

3   Materials & methods

Three multilevel observation wells were drilled, each located adjacent to a productive well. The position of the wells and piezometers screens at the two sites selected for the study are shown in Figure 7.2 (Jozefowo and Wladyslawowo). The well in Jozefowo is located in a region where groundwater extraction occurs, while the other is in Wladyslawowo and is located in a region where groundwater extraction does not occur. This situation enables the observation of groundwater chemistry changes in a vertical profile and enables the investigation of changes in groundwater chemistry with relation to the water exploitation condition.

Groundwater sampling was performed in December 2017. Samples were collected in 100-ml HDPE polyethylene bottles. Separate samples were taken for nutrient analyses (treated with chloroform) and for iron and manganese testing (treated with HNO_3). After sampling, water was stored in a portable refrigerator and immediately (the same day) transported to the laboratory. Water colour, electrical conductivity, alkalinity, pH and temperature were measured directly in the field. The chemical analyses were performed at Adam Mickiewicz University in Poznan (Institute of Geology) using a Compact IC 881Pro ionic chromatograph. As a quality control measure, the ionic error balance was calculated. The calculated error did not exceed 3%.

4   Results

The work shows a clear vertical groundwater differentiation at both investigated sites. The most clear differentiation is visible in case of nitrate. In Wladysawowo, a high nitrate concentration is observed only in the shallow part of the aquifer (W I). The nitrate concentration decrease with depth to less than 2 mg/l. In the shallow part chloride, sulfate and boron concentrations are higher than the deep part. In Józefowo relative high nitrate concentration is observed in both the shallow (J I) and deep part of the aquifer (well 1 bis, well 1a). Also chloride and sulfate concentrations are higher than in Wladyslawowo in both the shallow and deep part of the flow system.

5   Discussion

It was found previously (Dragon, 2013) that in a small area, relatively distinct groundwater chemistry differentiation occurs (Figure 7.3). The most distinct variation was found in nitrate concentration. The highest nitrate concentration occurred in the unconfined part of

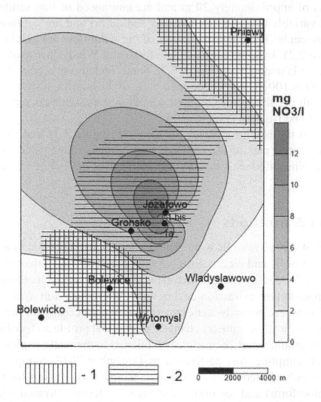

Figure 7.3 Changes in the hydrochemical parameters against a background of the nitrate concentrations (after Dragon, 2013, modified) 1 – total hardness > 6.0 mval/l; 2 – sulfate > 70 mg/l.

the aquifer. In the wells around this region, the concentrations are small, and they decrease along the flow path. Notably, nitrate appears in the central part of the recharge area, even in deep wells (>80 m). The total hardness and sulfate concentrations (products of the denitrification) are the lowest in the central part of the area, and they increase along the flow lines (Dragon, 2013).

The results of the current hydrochemical investigation are presented in Table 7.1 and on the Piper diagram (Figure 7.4).

The data show a different nature for the vertical groundwater chemistry changes observed in the two investigated sites (Table 7.1). It is apparent that most of the sampling points are clustered proximally on the piper diagram and that three sampling points differ from the others (Figure 7.4). These clustered points represent water from Jozefowo (from both: the shallow part of the aquifer – J I, and the deep part of the aquifer – J III and wells) and one point in Wladyslawowo representing the shallow part of the aquifer (W I). While the outliers (well Wladyslawowo 1 and W III) characterise groundwater from the deep part of the aquifer in Wladyslawowo. The sampling point W II is intermediate between the outliers and the clustered group. What is important is that groundwater chemistry in Jozefowo (where water extraction is performed) is very similar in the shallow and deep part of the aquifer at the two sites, except the nitrate concentration, which is notably higher in the shallow part.

Table 7.1 The results of hydrochemical investigation (sampling performed in December 2017)

Parameter	Unit	Władysławowo				Józefowo				
		W I	W II	W III	Well I	J I	J III	Well I bis	Well 2	Well 1a
pH	[-]	7.56	7.72	7.46	7.48	7.6	8.03	7.61	7.85	8.06
Alkalinity	mval/l	3.9	3.7	3.6	4.5	4.1	2.8	3.7	4.0	2.7
Total hardness		6.1	4.4	3.6	4.5	5.8	4.8	5.6	5.7	4.3
Electrical conductivity	µS/cm	730	500	462	462	593	525	560	592,0	566
B	µg/l	39.0	20.5	5.27	8.67	5.22	3.91	6.14	5.88	3.55
F	mg/l	0.21	0.16	0.15	0.14	0.23	0.27	0.21	0.20	0.22
Fe		0.16	0.39	1.90	1.68	0.06	0.07	0.08	0.41	0.21
Mn		0.05	0.06	0.11	0.12	0.04	0.04	0.06	0.06	0.04
Cl		25.6	14.0	3.10	3.29	25.1	19.9	19.8	18.2	16.2
NO$_3$		34.1	1.42	0.915	1.0	13.6	1.55	10.20	1.42	4.61
NO$_2$		0.012	0.009	0.007	0.011	0.020	0.015	0.045	0.007	0.025
NH$_4$		0.52	0.69	0.75	0.72	0.16	0.04	0.06	0.13	0.04
SO$_4$		62.2	28.6	9.79	10.4	55.6	82.9	72.8	72.3	60.8
Ca		105.1	75.0	60.2	75.8	98.7	81.0	95.1	96.4	71.4
Mg		10.9	7.89	7.50	8.94	11.0	9.21	10.2	10.8	8.58
Na		6.04	5.23	4.39	5.33	6.44	4.85	5.00	5.19	4.09
K		1.26	2.77	3.11	1.36	1.15	1.39	1.27	1.19	0.95
Screen interval	m	31.0–32.5	41.9–42.9	54.0–55.0	64.0–100.0	27.5–28.5	46.5–47.5	55.5–79.5	62.0–87.0	59.6–77.0

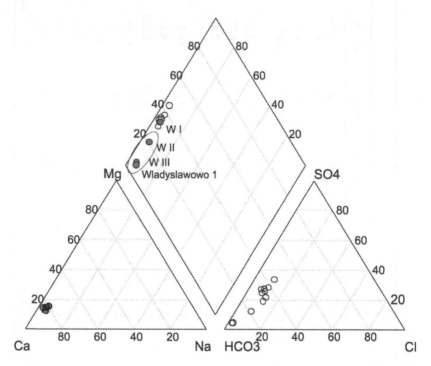

Figure 7.4 Piper diagram (sampling points signature according to Table 7.1)

In Wladyslawowo (no groundwater extraction), groundwater chemistry in the deep part of the aquifer differs considerably from that observed in the shallow part of the aquifer. The most distinct differentiation is visible in the case of nitrate (34 mg/l in the shallow part, in comparison to 1 mg/l in the deep part). There is also a large differentiation in sulfate, calcium and chloride concentrations. The concentrations of these constituents are much higher in the shallow part of the aquifer.

In productive wells located in Jozefowo there is also a relatively large differentiation of the nitrate concentrations detected in each well. The highest nitrate concentration (>10 m/l) is observed in well 1 bis, and the lowest (>1.5 mg/l) is observed in well 2. This can be caused by several factors. The most important is the low permeability of silty sand and till stratifications observed in the geological profile (Figure 7.2). It was recorded in piezometer J III and well 2. The position of the well screen is also important because the highest nitrate concentration was detected in the shallowest well (1 bis – Figure 7.2).

The nature of the groundwater chemistry indicates that the main contamination source is diffuse from agricultural use of both manure spreading and chemical fertilisers. This finding is supported by the high nitrate concentration in groundwater but relatively low concentration of chloride and boron (Gorski, 1989). The Wladyslawowo site is the only exception, because in the shallow part of the aquifer, boron concentrations were found to be relatively high.

The work confirms earlier findings related to groundwater flow system (Dragon, 2013). The conceptual model was formulated before the drilling of the observation wells. The

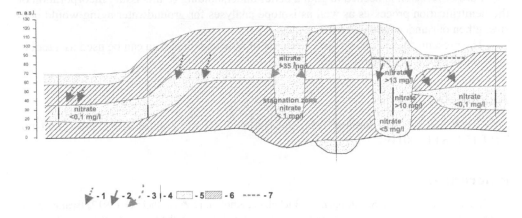

Figure 7.5 A conceptual model of the groundwater circulation and behaviour of the nitrate in the regional recharge zone of the Quaternary flow system (after Dragon, 2013 – modified according to current research). 1 – The preferential aquifer recharge through the aquitard, 2 – the preferential aquifer recharge – the unconfined parts of the flow system and the regions of intensive groundwater exploitation, 3 – the aquifer recharge under natural groundwater flow conditions (without exploitation), 4 – wells in regions of intensive groundwater exploitation, 5 – aquifers, 6 – aquitard, 7 – static and dynamic water level

obtained data confirms that the most intensive aquifer recharge occurs in the unconfined parts of the aquifer. Also, the groundwater occurring in the shallow part of the aquifer is highly contaminated (mainly by nitrate). In the region of groundwater extraction, these contaminants migrate downward to the deep part of the flow system. In the regions where natural gradients exist (without water extraction) the recharge in the shallow part of the aquifer can be intensive, but freshwater does not reach the deep part of the flow system. In this condition, a stagnation zone typically exists in the deep part of the aquifer (as defined by Toth, 1963), which is manifested by a completely different groundwater chemistry than in the shallow part of the aquifer.

6 Conclusions

The groundwater chemistry that occurs in the recharge zone of the Quaternary flow system (Lwówek region, Poland) is dependent on the groundwater flow system.

Contamination manifested by high nitrate concentrations detected in the shallow part of the aquifer can migrate to the deep part of the system because downward gradients (a characteristic attribute of the regional recharge zones) enable contaminants to move downward, especially in the regions of groundwater extraction. In these regions high nitrate concentration appears even at great depths (>80 m).

In the regions where natural gradients exist (without water extraction) the recharge in the shallow part of the aquifer is also intensive, but freshwater does not reach the deep part of the flow system. In this condition, the deep part of the aquifer typically exhibits a stagnation zone with different groundwater chemistry than that in the shallow part of the aquifer.

Further research is needed to gain a better understanding of this issue. Incorporation of the denitrification processes as well as isotope analyses for groundwater aging would help strengthen our understanding of this topic.

The presented research is a clear example of how chemical data can be used as a tool to investigate groundwater circulation.

Acknowledgments

This work has received funding from the National Science Centre of Poland (grant no. 2014/15/B/ST10/00119).

References

Al-Mashaikhi, K., Oswald, S., Attinger, S., Buchel, G., Knoller, K. & Strauch, G. (2012) Evaluation of groundwater dynamics and quality in the Najd aquifers located in the Sultanate of Oman. *Environmental Earth Science*, 66, 1195–1211.

Carillo-Rivera, J.J., Cardona, A. & Moss, D. (1996) Importance of the vertical component of groundwater flow: A hydrochemical approach in the valley of San Luis Potosi, Mexico. *Journal of Hydrology*, 185, 23–44.

Caschetto, M., Robertson, W., Petitta, M. & Aravena, R. (2018) Partial nitrification enhances natural attenuation of nitrogen in a septic system plume. *Science of the Total Environment*, 625, 801–808.

Chen, J., Tang, C., Sakura, Y., Yu, J. & Fukushima, Y. (2005) Nitrate pollution from agriculture in different hydrogeological zones of the regional groundwater flow system in the North China Plain. *Hydrogeology Journal*, 13, 481–492.

Coetsiers, M. & Walraevens, K. (2006) Chemical characterization of the Neogene Aquifer, Belgium. *Hydrogeology Journal*, 14, 1556–1568.

Czekaj, J., Jakóbczyk-Karpierz, S., Rubin, H., Sitek, S. & Witkowski, A.J. (2016) Identification of nitrate sources in groundwater and potential impact on drinking water reservoir (Goczałkowice reservoir, Poland). *Physics and Chemistry of the Earth*, 94, 35–46.

Dragon, K. (2013) Groundwater nitrate pollution in the recharge zone of a regional Quaternary flow system (Wielkopolska region, Poland). *Environmental Earth Science*, 68, 2099–2109.

Dragon, K. & Gorski, J. (2015) Identification of groundwater chemistry origins in a regional aquifer system (Wielkopolska region, Poland). *Environmental Earth Science*, 73(5), 2153–2167.

Dragon, K., Kasztelan, D., Gorski, J. & Najman, J. (2016) Influence of subsurface drainage systems on nitrate pollution of water supply aquifer (Tursko well-field, Poland). *Environmental Earth Science*, 75, 100.

Gorski, J. (1989) Główne problemy chemizmu wód podziemnych utworów kenozoiku środkowej Wielkopolski – in Polish. (The main hydrochemical problems of cainozoic aquifers located in Central Wielkopolska (Great Poland). *Zeszyty Nauk. AGH*, 45. Kraków.

Górski, J., Dragon, K. & Kaczmarek, P. (2019) Nitrate pollution in the Warta River (Poland) between 1958 and 2016: trends and causes. *Environmental Science and Pollution Research*, 26, 2038–2046.

Hendry, M.J. & Schwartz, F.W. (1990) The chemical evolution of groundwater in the Milk River Aquifer Canada. *Ground Water*, 28(2), 253–261.

Hudak, P.F. (2000) Regional trends in nitrate content of Texas groundwater. *Journal of Hydrology*, 228, 37–47.Ochsenkuhn, K.M., Kontoyannakos J. & Ochsenkuhn-Petropulu, M. (1997) A new approach to a hydrochemical study of groundwater flow. *Journal of Hydrology*, 194, 64–75.

Ophori, D.U. & Toth, J. (1989) Patterns of ground-water chemistry, Ross Creek Basin, Alberta, Canada. *Ground Water*, 27(1), 20–26.

Petitta, M., Fracchiolla, D., Aravena, R. & Barbieri, M. (2009) Application of isotopic and geochemical tools for the evaluation of nitrogen cycling in an agricultural basin, the Fucino Plain, Central Italy. *Journal of Hydrology*, 372, 124–135.

Rodvang, S.J. & Simpkins, W.W. (2001) Agricultural contaminants in Quaternary aquitards: A review of occurrence and fate in North America. *Hydrogeology Journal*, 9, 44–59.

Toth, J. (1963) A theretical analysis of ground water flow in small drainage basins. *Journal of Geophysical Research*, 68(16), 4795–4811.

Zurek, A., Rozanski, K., Mochalski, P. & Kuc, T. (2010) Assessment of denitrification rate in fissured-karstic aquifer near Opole (South-West Poland): combined use of gaseous and isotope tracers. *Biuletyn PIG*, 441, 209–216.

Chapter 8

Adsorption and desorption parameters of erythromycin migration in saturated porous media based on column tests

M. Okońska & K. Pietrewicz

list of symbols

α longitudinal dispersivity [L]
α_L Langmuir constant [L³/M]
β_L total sorption capacity of the solid phase [M/M]
C substance concentration in the liquid phase [M/L³]
C_0 substance concentration in the injected solution [M/L³]
C_u uniformity coefficient [-]
H water level [L]
i hydraulic gradient [L/L]
k hydraulic conductivity [L/T]
k_2 first reversible sorption rate coefficient [L³/MT]
k_3 second reversible sorption rate coefficient [1/T]
K_H Henry distribution coefficient [L³/M]
K_F Freundlich sorption coefficient [L³/M]
K_{OC} distribution coefficient [L³/M] with respect to the organic content OC [%]
L column length [L]
n_e effective porosity [$-$]
n_F Freundlich sorption exponent [$-$]
r correlation coefficient [-]
R retardation factor [-]
RMSE root mean square error [M/L³]
RR mass recovery [%]
Q volumetric flow rate [L³/T]
ρ_b bulk density of the porous medium [M/L³]
t time [T]
t_0 average time of water flow through the soil sample [T]
t_{in} time interval of the substance injection, when $C(t) = C_0$ at the input [T]
x distance [L]

1   Introduction

The literature provides an increasing amount of information on the presence of pharmaceuticals in surface water and groundwater. The occurrence of pharmaceuticals in water has been observed over the past dozen years or so (Hirsch *et al*., 1999; Mompelat *et al*., 2009; Zuccato *et al*., 2010; Kuczyńska, 2017). According to a report by the National Association of Clean

Water Agencies and the Association of Metropolitan Water Agencies (Snyder *et al.*, 2009) and a report by WHO (2012), research has detected only a small concentration of pharmaceuticals in water. However, the reports point to the need for monitoring the occurrence of such contaminating substances in the aquatic environment. It is necessary to determine the impact they exert on human health following long-term exposure, and also the possibilities and effects of their accumulation in the human body. Antibiotics and their combinations found in water pose a potential threat to aquatic ecosystems and the organisms living in them (Fent *et al.*, 2006; Kim and Aga, 2007; González-Pleiter *et al.*, 2013).

Erythromycin, a macrolide antibiotic used in veterinary practice and in human health care, is one of many pharmaceuticals that are found in water (e.g. Watkinson *et al.*, 2009; Yao *et al.*, 2017). It may cause many adverse effects, particularly with patients with liver or kidney deficiency. It may also interact with other drugs, decreasing their effectiveness or increasing their toxic impact on the human body (Ludden, 1985; Kessler *et al.*, 1986; Alomar, 2014; Ma *et al.*, 2014).

In the wastewater treatment process, the degree of erythromycin removal from the contaminated water varies and depends on the technologies applied (Watkinson *et al.*, 2007; Lin *et al.*, 2010; Al Qarni *et al.*, 2016). Under conventional treatment, by utilising such processes as dioxychlorination, sand filtration and ozonation, up to 95% of erythromycin can be removed from sewage, and up to 99% under advanced treatment (Boleda *et al.*, 2011). However, erythromycin can enter the water cycle along with untreated sewage or through washouts of solid waste or animal faeces. For those reasons, it is essential to examine the migration of erythromycin in various water-bearing media and in various conditions (Holmström *et al.*, 2003; De Voogt *et al.*, 2009; Ma *et al.*, 2015; Michael-Kordatou *et al.*, 2015). The sorption processes occurring during the migration of the antibiotic may delay its transport. Moreover, in the case of sorption processes, a hysteresis phenomenon can be observed. Compared to the adsorption process it features a different course of desorption, which occurs during the substance leaching from porous media (Limousin *et al.*, 2007). Research on the adsorption and leaching of erythromycin in porous media was conducted by a number of authors, including Ribeiro and Ribeiro (2003), Sun *et al.* (2009), Siemens *et al.* (2010) and Jin *et al.* (2014).

The aim of the presented research was to recognise erythromycin sorption behaviour during its migration through two different saturated porous media, determine the mathematical sorption model and estimate sorption parameters based on the observed breakthrough curves.

The findings will contribute to a better understanding of the antibiotic behaviour during its transport through saturated porous media, and the research results can be used in the mathematical modelling of the antibiotic adsorption and its leaching in the groundwater environment.

2   Materials and methods

2.1   Column tests

The research was conducted in laboratory conditions as column tests. A filtration column 50 cm long and 9 cm in diameter was filled with glass granules. Ball-shaped glass granules of 600–800 µm in diameter formed a homogeneous, isotropic porous medium. In the second experiment, the column was filled with a natural porous sediment – a sample of sand from Sierosław, located in central Wielkopolska region, Poland (Figure 8.1). The sand was medium quartz sand, less uniform in grain size compared to glass granules. The sand sample contained 0.114% of total organic carbon (TOC), and a trace amount of clay including such clay minerals as vermiculite, kaolinite and illite.

The porous media were fully saturated with distilled water. The flow through the sample was induced by the difference in water levels H_1 and H_2 and the constant hydraulic gradient i: 0.015 in the test with glass granules, and 0.021 in the test with sand. Firstly, in either experiment, a sodium chloride solution was injected. This substance was regarded as a conservative tracer (CTR), non-adsorbed, which is subjected to advection-dispersion processes and the breakthrough curve of which can be used to estimate the adsorbed substance migration (Leibundgut *et al.*, 2009; Zhao *et al.*, 2017). Following the leaching of the conservative tracer, an erythromycin solution (ERY) was injected.

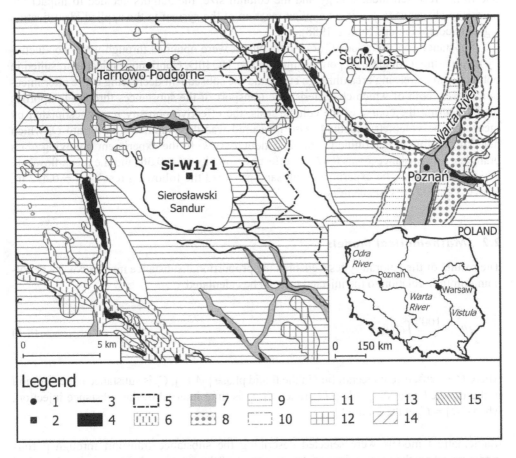

Figure 8.1 The localization of the sand sample intake point for the column test against the geomorphological forms of a part of the Poznań Lake District (on the basis of Karczewski *et al.*, 2007):
1 – settlement, 2 – sample point, 3 – river, 4 – lake, 5 – the border of the Poznań agglomerate, 6 – tunnel valley, 7 – flood plain, 8 – fill terrace, 9 – strath terrace, 10 – cut terrace, 11 – moraine upland, 12 – hillocks, 13 – sandur, 14 – eskers, 15 – thaw depressions.

For both samples a pulse injection followed by a continuous injection were conducted. In the case of sand, the tests involved the pulse injection of CTR and both pulse and continuous injections of ERY. The short-term injections were already described in an article by Okońska and Pietrewicz (2018). The continuous injections are the subject of this study. This kind of injection is regarded as an identification signal comprising: 1) the injection, during which the porous soil sample is saturated with a solution of known substance concentration C_0, and 2) the leaching of the substance and replacing it with water in which the substance concentration $C = 0$.

As erythromycin tends to photodegradate (Batchu *et al.*, 2014; Michael-Kordatou *et al.*, 2015), the experiments were conducted with limited access for sunlight to the laboratory.

The concentration of the examined substances in the injected solution was $C_0 = 500$ mg/dm³ in the case of sodium chloride, and $C_0 = 1541$ mg/dm³ in the case of erythromycin. In the literature, the occurrence of erythromycin in the aquatic environment is registered at lower concentration ranges (Peng *et al.*, 2014; Ma *et al.*, 2015; Yao *et al.*, 2017). However, due to the research methodology and the column size, the authors decided to impact the porous media so as to obtain an explicit reaction of the system to the applied identification signal (Söderström and Stoica, 1988).

Substance concentrations C at column output were obtained through water conductivity measurements made with portable multi meter HQ40D. Continuous injection resulted in obtaining an increasing breakthrough curve and the subsequent decreasing curve after injection ceases, both of which characterise the transport of the substances through the porous media. During the course of the column tests, the effective porosity n_e was estimated by means of the volumetric method. Also measured was the volumetric flow rate Q, which made it possible to determine hydraulic conductivity k. In the experiment involving glass granules, flow distance x was 0.470 m, effective porosity $n_e = 0.36$, and the mean flow rate $Q = 2.4E-07$ m³/s. The experiment with sand provided the following results: $x = 0.472$ m, $n_e = 0.33$ and $Q = 1.4E-08$ m³/s.

2.2   Mathematical models

The results of the laboratory tests enabled the proportional substance mass recovery at column output in relation to the injected mass to be calculated:

$$RR = \frac{100}{C_0 t_{in}} \int_0^t C(t)\,dt \qquad (8.1)$$

where C is substance concentration in the liquid phase [M/L³], C_0 is substance concentration in the injected solution [M/L³], t is time [T], t_{in} is time interval of the substance injection, when $C(t) = C_0$ at the input [T].

Mathematical models were selected describing the substance transport through porous media based on the mass balances. The transport of the conservative tracer was described with a one-dimensional equation that takes into account the advection-dispersion processes (A-D model) (Weber *et al.*, 1991). In the mathematical model, it was assumed that the influence of molecular diffusion on the substance migration through the media is small and can be neglected, particularly in the case of glass granules (Okońska and Pietrewicz, 2018). As regards erythromycin, seven out of eleven mathematical models describing the advection, dispersion and sorption processes were considered (their mathematical equations can be found in Okońska et al., 2017). The sorption part accounted for the following:

- equilibrium sorption as one of the commonly applied models: the Henry model (H model), the Freundlich model (F model), or the Langmuir model (L model); or
- non-equilibrium sorption: reversible sorption (R model); or
- the simultaneous coexistence of equilibrium and non-equilibrium sorption, described by the following set of equations: the Henry model with reversible sorption (H-R model), the Freundlich model with reversible sorption (F-R model), or the Langmuir model with reversible sorption (L-R model).

The models consisting of only one sorption type are called simple sorption models, while the models that account for the coexistence of two sorption types are called hybrid (two-site) models. The mathematical models neglected the chemical and bio-chemical reactions, biodegradation, chemo- and electro-osmotic and capillary effects. Furthermore, it was assumed that the porous material is undeformable and fully saturated with liquid solution, and that all the material and geometrical parameters present in the model are constant.

When solving differential equations, the authors assumed appropriate boundary conditions for the continuous injection:

- The initial condition, determining the distribution of concentrations in all of the analysed area, as: $C(x,t)\big|_{x\geq0,t=0} = 0$, where C is substance concentration in the liquid phase [M/L³], x is distance [L] and t is time [T].
- The first-type (Dirichlet) boundary condition at column input: $C(x,t)\big|_{x=0,t>0} = C_0$, where C_0 is substance concentration in the injected solution [M/L³].
- The second-type (Neumann) boundary condition at column output: $\dfrac{\partial C(x,t)}{\partial t}\bigg|_{x=L,t>0} \in 0, C_0$, where L is the column length [L].

As regards the leaching of the substance from the porous media, the authors assumed as follows:

- The initial condition: $C(x,t)\big|_{x\geq0,\ t=0} = C_0$.
- The first-type (Dirichlet) boundary condition at column input: $C(x,t)\big|_{x=0,t>0} = 0$.
- The second-type (Neumann) boundary condition at column output: $\dfrac{\partial C(x,t)}{\partial t}\bigg|_{x=L,t>0} \in C_0, 0$.

2.3 Interpretation

The numerical solutions of the differential equations of transport and sorption were conducted in the computing environment MATLAB utilising numerical optimisation methods. The identification procedure was launched after the data from the column test and the initial values of the migration parameters had been entered. The identification involved solving a mathematical model by means of the *pdepe* function – a solver utilising the finite element method (FEM), and then calculating the error function, and comparing the result against the optimisation criteria. For this purpose, the authors applied the *Optimization Toolbox* and the built-in *lsqcurvefit* function dedicated to the non-linear solution of the optimisation issue. The iterative process was conducted for the corrected parameter values, until the set optimisation criteria (the defined arguments of the *optimset* function) were met.

In the A-D model, the authors searched for the value of longitudinal dispersivity α, and specified the value of hydraulic conductivity k with a possible measurement error ±10%. In the ERY migration models, several sorption parameters were identified. The values of longitudinal dispersivity α and hydraulic conductivity k were substituted with the parameter values set for CTR.

The convergence between theoretical and experimental breakthrough curves was analysed qualitatively by calculating the root mean square error RMSE and correlation coefficient r (Okońska *et al.*, 2017).

The sorption capacity of erythromycin was determined on the basis of retardation factor R (Witczak *et al.*, 2013; Kret *et al.*, 2015). First, it was estimated by comparing the experimental breakthrough curves of CTR and ERY (Małecki *et al.*, 2006; Sieczka *et al.*, 2018). Next, the value of the retardation factor R [-] was calculated on the basis of the Henry distribution coefficient K_H [L³/M] (Okońska, 2006; Kret *et al.*, 2015):

$$R = 1 + \frac{\rho_b}{n_e} K_H \tag{8.2}$$

where ρ_b is the bulk density of the porous medium [M/L³].

Moreover, on the basis of distribution coefficient K_H, the authors calculated the normalised distribution coefficient K_{OC} [L³/M] with respect to the organic content OC [%] of the porous medium (Appelo and Postma, 1999):

$$K_{OC} = K_H \frac{100}{OC} \tag{8.3}$$

3   Results

3.1   Experimental breakthrough curves

The breakthrough curves registered during the laboratory tests are shown in Figure 8.2. The charts present the concentration of substance C, measured at column output, in relation to concentration C_0 in the injected solution. Meanwhile, the time of the substance migration through the column was referred to the average time of water flow through sample t_0 [T], which depends on the flow rate Q [L³/T]:

$$t_0 = \frac{n_e \pi d^2 x}{4Q} \tag{8.4}$$

A comparison of the impulse breakthrough curves of CTR and ERY (Figures 8.2IA and 8.2IIA) shows the retardation of the antibiotic migration in relation to the non-adsorbed tracer. Retardation factor R, calculated for those breakthrough curves, was 1.1 for glass granules and 1.4 for sand, which points to a small sorption capacity of erythromycin.

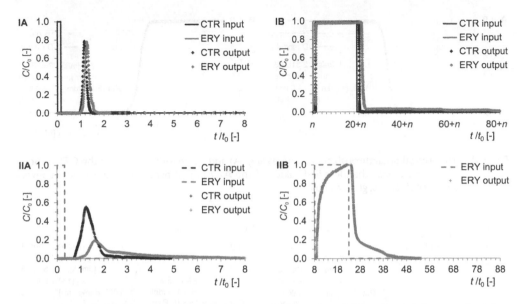

Figure 8.2 Breakthrough curves through glass granules (I) and sand (II) registered during pulse injection (A) and continuous injection (B): *n* signifies value 36.18 for CTR and 41.63 for ERY.

In the tests utilising glass granules, 100% CTR mass and 99% ERY mass was recovered (±5%, Equation 1). In the test utilising natural sediment, mass recovery of sodium chloride was slightly more than 100%, possibly caused by the type of the conservative substance used in the tests and the applied method of measuring concentration at column output. Sodium ions could have been leached slightly more slowly than chloride ions, thus having a small influence on the shape of the breakthrough curve in its lower register. Erythromycin mass recovery in the tests utilising sand was 96%. Allowing for some uncertainty of calculating mass balance ±5%, it can be decided that in the test with natural sediment 90–100% of erythromycin mass was leached from the medium. In further considerations, the authors applied a mathematical model, based on the test with glass granules, that assumed no loss of the migrating substance mass following adsorption on the sediment or photodegradation.

3.2 The estimation of migration parameters

The theoretical pulse breakthrough curves of CTR and ERY through glass granules and sand, along with the identified values of migration parameters are described by Okońska and Pietrewicz (2018). Figures 8.3–8.5 show the increasing and the decreasing (leaching) breakthrough curves of CTR and ERY calculated in the MATLAB environment.

The numerical solution of the advection-dispersion equation describing the transport of CTR (A-D model) for the glass granules, showed the value of longitudinal dispersivity α equal to be 0.0008 m for the increasing breakthrough curve and 0.0010 m for the decreasing breakthrough curve (Figure 8.3). In the case of sand, the value of longitudinal dispersivity α, determined on the basis of the pulse breakthrough curve, was 0.0096 m (RMSE = 0.036 mg/dm³ and $r = 0.997$).

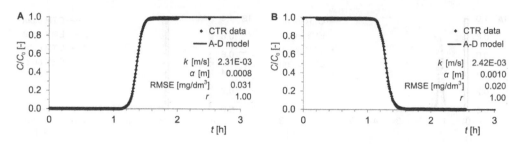

Figure 8.3 Theoretical breakthrough curves, increasing (A) and decreasing (B), against the CTR experimental data, with the identified values of the A-D model parameters and model convergence coefficients, glass granules.

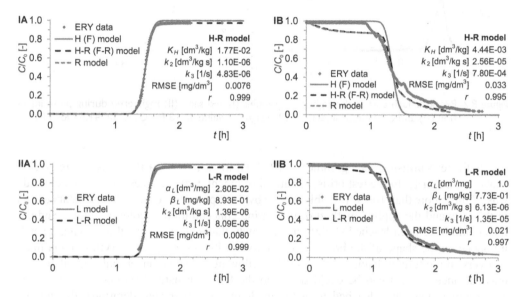

Figure 8.4 Theoretical breakthrough curves, increasing (A) and decreasing (B), against ERY experimental data, with the identified parameter values of the selected mathematical models and model convergence coefficients, glass granules.

For the two different porous media and the same distance, various values of longitudinal dispersivity were obtained. The CTR that migrated through sand was more dispersed. Sand, as opposed to glass granules, is characterised by lower uniformity of grain size and lower porosity. The uniformity coefficient C_u (Holtz and Kovacs, 1981) and porosity n_e are as follows: for sand $C_u = 4.13$ and $n_e = 0.33$, and for glass granules $C_u = 1.20$ and $n_e = 0.36$. During the course of the experiments, higher dispersivity was not observed to correlate with a higher value of hydraulic conductivity k. Klotz *et al.* (1980), Xu and Eckstein (1997), Arriaza and Ghezzehei (2013), Zhao *et al.* (2017) argue that dispersivity depends on a number of factors, including the physical properties of the porous medium such as uniformity of grain size distribution, mean particle size, particle shape, flow velocity, or porosity. The obtained results are in line with the observations by Xu and Eckstein (1997), whose column tests did

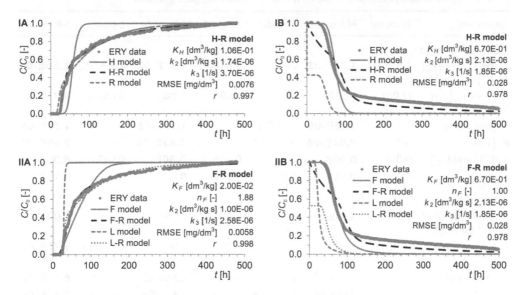

Figure 8.5 Theoretical breakthrough curves, increasing (A) and decreasing (B), against ERY experimental data, with the identified parameter values of the selected mathematical models and model convergence coefficients, sand.

not confirm that dispersivity was related to the average hydraulic conductivity. However, they pointed to the fact that dispersivity values increase with the decrease of porosity, and that dispersivity depends mainly on the uniformity of grain size, particularly regarding material of uniformity coefficient C_u higher than 3 (see also Sieczka and Koda, 2016).

An analysis of the value of the dispersivity, depending on the type of the breakthrough curve, shows that in the case of the glass granules the differences are small – during the leaching process, dispersivity was only slightly higher than in the case of the pulse breakthrough curve ($\alpha = 0.0007$m, RMSE $= 0.014$ mg/dm^3 and $r = 0.999$) and the increasing breakthrough curve ($\alpha = 0.0008$ m, RMSE $= 0.031$ mg/dm^3 and $r = 1.000$).

Based on the results of the model tests of erythromycin migration and the calculated convergence coefficients RMSE and r, both for glass granules and sand, the highest convergence between theoretical breakthrough curves and the experimental ones was obtained for hybrid sorption models.

In the case of the glass granules, the F-R model showed that numerical erythromycin breakthrough curves converged with the ones set for the H-R model (Figure 8.4). The RMSE and r coefficients, obtained through optimisation, show that erythromycin adsorption on glass granules, i.e. the increasing breakthrough curve, is well described by both the H-R model (and F-R model) as well as the L-R model; while in terms of the leaching breakthrough curve, the highest convergence level was obtained for the L-R model. The values of the parameters of all mathematical sorption models and the indicators of the convergence between theoretical breakthrough curves and the experimental ones are presented in Table 8.1 (glass granules) and Table 8.2 (sand).

Erythromycin adsorption on a natural medium is best described by the F-R model. A high convergence between the theoretical increasing breakthrough curve and the experimental

Table 8.1 The parameter values of mathematical sorption models identified during the course of the optimisation procedure in the MATLAB environment, glass granules

Parameters	H model	H-R model	R model	F model	F-R model	L model	L-R model
Continuous curve							
K_H or K_F [dm³/kg] or α_L [dm³/mg]*	1.83E-02	**1.77E-02**	–	1.83E-02	**1.77E-02**	2.99E-02	**2.80E-02**
n_F [-] or β_L [mg/kg]*	(1.00)	**(1.00)**	–	1.00	**1.00**	8.86E-01	**8.93E-01**
k_2 [dm³/kg s]	–	**1.10E-06**	5.20E-04	–	**1.10E-06**	–	**1.39E-06**
k_3 [1/s]	–	**4.83E-06**	2.81E-02	–	**4.83E-06**	–	**8.09E-06**
RMSE [mg/dm³]	0.019	**0.008**	0.021	0.019	**0.008**	0.022	**0.008**
r [-]	0.999	**0.999**	0.997	0.999	**0.999**	0.999	**0.999**
Leaching curve							
K_H or K_F [dm³/kg] or α_L [dm³/mg]*	7.31E-03	4.44E-03	–	7.31E-03	4.44E-03	1.00	**1.00**
n_F [-] or β_L [mg/kg]*	(1.00)	(1.00)	–	1.00	1.00	6.38E-01	**7.73E-01**
k_2 [dm³/kg s]	–	2.56E-05	2.90E-05	–	2.56E-05	–	**6.13E-06**
k_3 [1/s]	–	7.80E-04	9.25E-04	–	7.80E-04	–	**1.35E-05**
RMSE [mg/dm³]	0.058	0.033	0.034	0.058	0.033	0.031	**0.021**
r [-]	0.983	0.995	0.996	0.983	0.995	0.995	**0.997**

Note: *K_H for H and H-R models, K_F and n_F for F and F-R models, α_L and β_L for L and L-R models. For each curve, the parameters of the model with the lowest value of RMSE and the highest value of r are in bold

Table 8.2 The parameter values of mathematical sorption models identified during the course of the optimisation procedure in the MATLAB environment, sand

Parameters	H model	H-R model	R model	F model	F-R model	L model	L-R model
Continuous curve							
K_H or K_F [dm³/kg] or α_L [dm³/mg]*	2.97E-01	1.06E-01	–	2.81E-02	**2.00E-02**	9.79E-01	2.15E-01
n_F [-] or β_L [mg/kg]*	(1.00)	(1.00)	–	2.00	**1.88**	1.00	1.00
k_2 [dm³/kg s]	–	1.74E-06	3.07E-06	–	**1.00E-06**	–	2.30E-06
k_3 [1/s]	–	3.70E-06	5.85E-06	–	**2.58E-06**	–	4.53E-06
RMSE [mg/dm³]	0.050	0.008	0.015	0.030	**0.006**	0.074	0.010
r [-]	0.930	0.997	0.988	0.974	**0.998**	0.813	0.995
Leaching curve							
K_H or K_F [dm³/kg] or α_L [dm³/mg]*	4.49E-01	**6.70E-01**	–	1.00	**6.70E-01**	6.25E-01	4.42E-01
n_F [-] or β_L [mg/kg]*	(1.00)	**(1.00)**	–	0.74	**1.00**	1.00	1.00
k_2 [dm³/kg s]	–	2.13E-06	2.05E-04	–	**2.13E-06**	–	1.25E-01
k_3 [1/s]	–	**1.85E-06**	8.22E-04	–	**1.85E-06**	–	7.03E-01
RMSE [mg/dm³]	0.043	**0.028**	0.091	0.036	**0.028**	0.091	0.082
r [-]	0.960	**0.978**	0.892	0.974	**0.978**	0.739	0.920

Note: *K_H for H and H-R models, K_F and n_F for F and F-R models, α_L and β_L for L and L-R models. For each curve, the parameters of the model with the lowest value of RMSE and the highest value of r are in bold

data was also obtained for the H-R model (Figure 8.5). The same mathematical models were also selected in the case of the leaching breakthrough curve.

The results of the experimental and model tests show that erythromycin migration was different in both porous media. As long as the adsorption process on glass granules could be roughly described even with a simple sorption model, in terms of sand, the process must be described with the hybrid model (Brusseau et al., 1997). At the same time, with a view to obtain a good approximation, the H-R model should be utilised. The increasing breakthrough curve of the antibiotic migration through a natural and less homogeneous sediment is notably asymmetric. The set values of distribution coefficient K_H show that erythromycin is subject to more adsorption on the sand medium. Sand causes more retardation of the antibiotic migration through a natural medium than in the case of glass granules – according to Equation 2, retardation factor R is 1.57 and 1.08, respectively. On the basis of the mathematical models, it can be assumed that the course of adsorption/desorption process on the tested sand medium is close to a linear process. However, the adsorption process may recede as the sorption complex is being filled in. It seems that in the case of glass granules the maximum capacity of the sorption – reflected by the equilibrium element of the model – can play a major role for the adsorption/desorption process.

The leaching breakthrough curves follow a different course than the increasing breakthrough curves in both samples: they are characterised by a long "tail" which proves that the desorption process is slow (Limousin et al., 2007).

It is difficult to make a comparison of the observed values of sorption parameters with other results. Such tests are often carried out as batch tests, for different porous media, different substance concentrations, at different flow rates, or distance. Erythromycin sorption may use Langmuir and Freundlich models as the most appropriate to describe the course of the process in porous media. Comparison of the value of the predicted distribution coefficient log K_{OC} for erythromycin with the American Chemical Society (Siemens et al., 2010) and the parameter values calculated in this study for sand and the increasing curve (Equation 3), the results were similar, i.e.: 2.53 dm³/kg and 1.97 dm³/kg, respectively.

4   Conclusions

The numerical solution of the equations and the optimisation procedures that were used made it possible to estimate parameters values in multiparameter mathematical models.

On the basis of the CTR breakthrough curve, according to the advection-dispersion model, longitudinal dispersivity α was calculated, and showed the relationship between dispersivity and the uniformity and porosity. As for the increasing breakthrough curve for glass granules – an isotropic medium – dispersivity was 0.0008 m, while during the process of leaching from the medium, the dispersivity value was slightly higher and reached 0.0010 m. In the sand, which is characterised by lower uniformity and smaller porosity, the calculated dispersivity was 0.0096 m.

For both glass granules and sand, the best convergence between theoretical breakthrough curves of ERY and the experimental ones was obtained for hybrid sorption models, which combine equilibrium and non-equilibrium sorption.

As a result of applying the optimisation procedure in the MATLAB environment, the following adsorption parameters were obtained for glass granules (H-R model): K_H = 1.77E-02 dm³/kg, k_2 = 1.10E-06 dm³/kg s, and k_3 = 4.83E-06 1/s, with model convergence coefficients RMSE = 0.008 mg/dm³ and r = 1.00. Desorption parameters, estimated on the basis of the leaching breakthrough curve equal (L-R model): $\alpha_L = 1.0$ dm/mg, $\beta_L = 7.73E\text{-}01$ mg/kg, $k_2 = 6.13E\text{-}06$ dm³/kg s, and $k_3 = 1.35E\text{-}05$ 1/s (RMSE $= 0.021$ mg/dm³ and $r = 1.00$).

Sand adsorption parameters are as follows (F-R model): $K_F = 2.00E\text{-}02$ dm³/kg, $n_F = 1.88$, $k_2 = 1.00E\text{-}06$ dm³/kg s, and $k_3 = 2.58E\text{-}06$ 1/s (RMSE $= 0.006$ mg/dm³ and $r = 1.00$). Desorption parameters equal (H-R model): $K_H = 6.70E\text{-}01$ dm³/kg, $k_2 = 2.13E\text{-}06$ dm³/kg s, and $k_3 = 1.85E\text{-}06$ 1/s (RMSE $= 0.028$ mg/dm³ and $r = 0.98$).

The leaching breakthrough curves are characterised by a noticeable "tail", which confirms the slow desorption of erythromycin. A long observation of the substance leaching from the medium is essential. It allows a correct calculation of the substance mass balance and, subsequently, identifying an appropriate migration model. As regards further tests on erythromycin migration it is recommended that, for either medium, all types of non-adsorbed tracer breakthrough curves are recorded – not only the pulse one, but also the increasing and decreasing ones.

The calculated values of retardation factor R, on the basis of the registered breakthrough curves of erythromycin migration and the estimated values of the distribution coefficient (1.1 for glass granules, and 1.4–1.6 for sand), confirm the small capacity of the antibiotic adsorption on the investigated porous media.

The obtained results can be used in order to estimate erythromycin transport time in aquifers composed of sand with relatively low organic content and subsequently assess aquifer vulnerability to contamination.

Acknowledgements

The results presented in this study were obtained as part of research project no. DEC-2011/01/B/ST10/02063 financed by the National Science Centre resources (Poland).

References

Alomar, M.J. (2014) Factors affecting the development of adverse drug reactions (Review article). *Saudi Pharmaceutical Journal.* [Online], 22, 83–94. doi:10.1016/j.jsps.2013.02.003.

Al Qarni, H., Collier, P., O'Keeffe, J. & Akunna, J. (2016) Investigating the removal of some pharmaceutical compounds in hospital wastewater treatment plants operating in Saudi Arabia. *Environmental Science and Pollution Research.* [Online], 23, 13003–13014. doi:10.1007/s11356-016-6389-7.

Appelo, C.A.J. & Postma, D. (1999) *Geochemistry, Groundwater and Pollution.* Brookfield, A.A. Balkema, Rotterdam.

Arriaza, J.L. & Ghezzehei, T.A. (2013) Explaining longitudinal hydrodynamic dispersion using variance of pore size distribution. *Journal of Porous Media.* [Online], 16, 11–19. doi:10.1615/JPorMedia.v16.i1.20.

Batchu, S.R., Panditi, V.R., O'Shea, K.E. & Gardinali, P.R. (2014) Photodegradation of antibiotics under simulated solar radiation: Implications for their environmental fate. *Science of the Total Environment.* [Online], 470–471, 299–310. doi:10.1016/j.scitotenv.2013.09.057.

Boleda, M.R., Galceran, M.T. & Ventura, F. (2011) Behavior of pharmaceuticals and drugs of abuse in a drinking water treatment plant (DWTP) using combined conventional and ultrafiltration and reverse osmosis (UF/RO) treatments. *Environmental Pollution.* [Online], 159, 1584–1591. doi:10.1016/j.envpol.2011.02.051.

Brusseau, M.L., Hu, Q. & Srivastava, R. (1997) Using flow interruption to identify factors causing nonideal contaminant transport. *Journal of Contaminant Hydrology.* [Online], 24, 205–219. doi:10.1016/S0169-7722(96)00009-5.

De Voogt, P., Janex-Habibi, M.L., Sacher, F., Puijker, L. & Mons, M. (2009) Development of a common priority list of pharmaceuticals relevant for the water cycle. *Water Science & Technology.* [Online], 59, 39–46. doi:10.2166/wst.2009.764.

Fent, K., Weston, A.A. & Caminada, D. (2006) Ecotoxicology of human pharmaceuticals. *Aquatic Toxicology.* [Online], 76, 122–159. doi:10.1016/j.aquatox.2005.09.009.

González-Pleiter, M., Gonzalo, S., Rodea-Palomares, I., Leganés, F., Rosal, R., Boltes, K., Marco, E. & Fernández-Piñas, F. (2013) Toxicity of five antibiotics and their mixtures towards photosynthetic aquatic organisms: Implications for environmental risk assessment. *Water Research.* [Online], 47, 2050–2064. doi:10.1016/j.watres.2013.01.020.

Hirsch, R., Ternes, T., Haberer, K. & Kratz, K.L. (1999) Occurrence of antibiotics in the aquatic environment. *Science of the Total Environment.* [Online], 225, 109–118. doi:10.1016/S0048-9697(98)00337-4.

Holmström, K., Gräslund, S., Wahlström, A., Poungshompoo, S., Bengtsson, B.E. & Kautsky, N.L. (2003) Antibiotic use in shrimp farming and implications for environmental impacts and human health. *International Journal of Food Science & Technology.* [Online], 38, 255–266. doi:10.1046/j.1365-2621.2003.00671.x.

Holtz, R. & Kovacs, W. (1981) *An Introduction to Geotechnical Engineering.* Prentice-Hall, Englewood Cliffs, NJ.

Jin, X., Chen, K., Zhu, J.W. & Wu, Y.Y. (2014) Effect of solution polarity and temperature on adsorption separation of erythromycin A and C onto macroporous resin SP825. *Separation Science and Technology.* [Online], 49, 898–906. doi:10.1080/01496395.2013.863341.

Karczewski, A., Mazurek, M., Stach, A. & Zwoliński, Z. (2007) *Geomorphological Map of the Wielkopolska-Kujawy Lowland 1:300 000,* B. Krygowski (ed.). Numerical version. Instytut Paleogeografii i Geoekologii, Uniwersytet im. A. Mickiewicza, Poznań (in Polish).

Kessler, M., Louis, J., Renoult, E., Vigneron, B. & Netter, P. (1986) Interaction between cyclosporin and erythromycin in a kidney transplant patient. *European Journal of Clinical Pharmacology.* [Online], 30, 633–634. doi:10.1007/BF00542427.

Kim, S. & Aga, D.S. (2007) Potential ecological and human health impacts of antibiotics and antibiotic-resistant bacteria from wastewater treatment plants. *Journal of Toxicology and Environmental Health, Part B: Critical Reviews.* [Online], 10, 559–573. doi:10.1080/15287390600975137.

Klotz, D., Seiler, K.P., Moser, H. & Neumaier, F. (1980) Dispersivity and velocity relationship from laboratory and field experiments. *Journal of Hydrology.* [Online], 45, 169–184. doi:10.1016/0022-1694(80)90018-9.

Kret, E., Kiecak, A., Malina, G., Nijenhuis, I. & Postawa, A. (2015) Identification of TCE and PCE sorption and biodegradation parameters in a sandy aquifer for fate and transport modelling: batch and column studies. *Environmental Science and Pollution Research.* [Online], 22, 9877–9888. doi:10.1007/s11356-015-4156-9.

Kuczyńska, A. (2017) Results of a pilot study on the assessment of pharmaceuticals in groundwater in samples collected from the national groundwater monitoring network. *Przegląd Geologiczny,* 65(11/1), 1096–1103 (in Polish).

Leibundgut, C., Maloszewski, P. & Külls, C. (2009) *Tracers in Hydrology.* John Wiley & Sons Ltd., Chichester.

Limousin, G., Gaudet, J.P., Charlet, L., Szenknect, S., Barthes, V. & Krimissa, M. (2007) Sorption isotherms: A review on physical bases, modeling and measurement. *Applied Geochemistry.* [Online], 22, 249–275. doi:10.1016/j.apgeochem.2006.09.010.

Lin, A.Y.C., Lin, C.F., Tsai, Y.T., Lin, H.H.H., Chen, J., Wang, X.H. & Yu, T.H. (2010) Fate of selected pharmaceuticals and personal care products after secondary wastewater treatment processes in Taiwan. *Water Science & Technology.* [Online], 62, 2450–2458. doi:10.2166/wst.2010.476.

Ludden, T.M. (1985) Pharmacokinetic interactions of the macrolide antibiotics. *Clinical Pharmacokinetics.* [Online], 10, 63–79. doi:10.2165/00003088-198510010-00003.

Ma, T.K.W., Chow, K.M., Choy, A.S.M., Kwan, B.C.H., Szeto, C.C. & Li, P.K.T. (2014) Clinical manifestation of macrolide antibiotic toxicity in CKD and dialysis patients. *Clinical Kidney Journal.* [Online], 7, 507–512. doi:10.1093/ckj/sfu098.

Ma, Y., Li, M., Wu, M., Li, Z. & Liu, X. (2015) Occurrences and regional distributions of 20 antibiotics in water bodies during groundwater recharge. *Science of the Total Environment*. [Online], 518–519, 498–506. doi:10.1016/j.scitotenv.2015.02.100.

Małecki, J.J., Nawalany, M., Witczak, S. & Gruszczyński, T. (2006) *Determination of Contaminant Migration Parameters in a Porous Medium for Hydrogeological and Environmental Protection Research: Methodical Guide*. Ministerstwo Środowiska, Warszawa (in Polish).

Michael-Kordatou, I., Iacovou, M., Frontistis, Z., Hapeshi, E., Dionysiou, D.D. & Fatt-Kassinos, D. (2015) Erythromycin oxidation and ERY-resistant Escherichia coli inactivation in urban wastewater by sulfate radical-based oxidation process under UV-C irradiation. *Water Research*. [Online], 85, 346–358. doi:10.1016/j.watres.2015.08.050.

Mompelat, S., Le Bot, B. & Thomas, O. (2009) Occurrence and fate of pharmaceutical products and by-products, from resource to drinking water. *Environment International*. [Online], 35, 803–814. doi:10.1016/j.envint.2008.10.008.

Okońska, M. (2006) *The Identification of Pollutants Migration Parameters in a Groundwater Porous Medium by the Method of the Column Experiment Modelling*. Geologos 9, Bogucki Wydawnictwo Naukowe, Poznań (in Polish).

Okońska, M. & Pietrewicz, K. (2018) Identification of mathematical model and parameter estimation of erythromycin migration in two different porous media based on column tests. *Geologia Croatica*. [Online], 71(2), 47–53. doi:10.4154/gc.2018.05.

Okońska, M., Kaczmarek, M., Małoszewski, P. & Marciniak, M. (2017) The verification of the estimation of transport and sorption parameters in the MATLAB environment: Column test. *Geology, Geophysics & Environment*. [Online], 43(3), 213–227. doi:10.7494/geol.2017.43.3.213.

Peng, X., Ou, W., Wang, C., Wang, Z., Huang, Q., Jin, J. & Tan, J. (2014) Occurrence and ecological potential of pharmaceuticals and personal care products in groundwater and reservoirs in the vicinity of municipal landfills in China. *Science of the Total Environment*. [Online], 490, 889–898. doi:10.1016/j.scitotenv.2014.05.068.

Ribeiro, M.H.L. & Ribeiro, I.A.C. (2003) Modelling the adsorption kinetics of erythromycin onto neutral and anionic resins. *Bioprocess and Biosystems Engineering*. [Online], 26, 49–55. doi:10.1007/s00449-003-0324-2.

Sieczka, A., Bujakowski, F., Falkowski, T. & Koda, E. (2018) Morphogenesis of a floodplain as a criterion for assessing the susceptibility to water pollution in an agriculturally rich valley of a lowland river. *Water*. [Online], 10, 399–420. doi:10.3390/w10040399.

Sieczka, A. & Koda, E. (2016) Kinetic and equilibrium studies of sorption of ammonium in the soil-water environment in agricultural areas of central Poland. *Applied Sciences*. [Online], 6, 269. doi:10.3390/app6100269.

Siemens, J., Huschek, G., Walshe, G., Siebe, C., Kasteel, R., Wulf, S., Clemens, J. & Kaupenjohann, M. (2010) Transport of pharmaceuticals in columns of a wastewater-irrigated Mexican clay soil. *Journal of Environmental Quality*. [Online], 39, 1201–1210. doi:10.2134/jeq2009.0105.

Snyder, S., Lue-Hing, C., Cotruvo, J., Drewes, J.E., Eaton, A., Pleus, R.C. & Schlenk, D. (2009) *Pharmaceuticals in the Water Environment*. National Association of Clean Water Agencies, Association of Metropolitan Water Agencies. [Online]. Available from: www.nacwa.org/news-publications/white-papers-publications [Accessed 2 March 2018].

Söderström, T. & Stoica, P. (1988) *Systems Identification*. Prentice-Hall, Upper Saddle River, NJ.

Sun, Y., Zhu, J.W., Chen, K. & Xu, J. (2009) Modeling erythromycin adsorption to the macroporous resin Sepabead SP825. *Chemical Engineering Communications*. [Online], 196, 906–916. doi:10.1080/00986440902743802.

Watkinson, A.J., Murby, E.J. & Costanzo, S.D. (2007) Removal of antibiotics in conventional and advanced wastewater treatment: Implications for environmental discharge and wastewater recycling. *Water Research*. [Online], 41, 4164–4176. doi:10.1016/j.watres.2007.04.005.

Watkinson, A.J., Murby, E.J., Kolpin, D.W. & Costanzo, S.D. (2009) The occurrence of antibiotics in an urban watershed: From wastewater to drinking water. *Science of the Total Environment*. [Online], 407, 2711–2723. doi:10.1016/j.scitotenv.2008.11.059.

Weber, W.J., McGinley, P.M. & Katz, L.E. (1991) Sorption phenomena in subsurface systems: Concepts, models and effects on contaminant fate and transport. *Water Research*. [Online], 25, 499–528. doi:10.1016/0043-1354(91)90125-A.

Witczak, S., Kania, J. & Kmiecik, E. (2013) *Catalog of Selected Physical and Chemical Indicators of Groundwater Pollution and Methods of Their Determination*. Inspekcja Ochrony Środowiska, Warszawa (in Polish).

World Health Organization (2012) *Pharmaceuticals in Drinking-Water*. [Online]. Available from: www.who.int/water_sanitation_health/publications/2012/pharmaceuticals/en/ [Accessed 2 March 2018].

Xu, M. & Eckstein, Y. (1997) Statistical analysis of the relationships between dispersivity and other physical properties of porous media. *Hydrogeology Journal*. [Online], 5(4), 4–20. doi:10.1007/s100400050254.

Yao, L., Wang, Y., Tong, L., Deng, Y., Li, Y., Gan, Y., Guo, W., Dong, C., Duan, Y. & Zhao, K. (2017) Occurrence and risk assessment of antibiotics in surface water and groundwater from different depths of aquifers: A case study at Jianghan Plain, central China. *Ecotoxicology and Environmental Safety*. [Online], 135, 236–242. doi:10.1016/j.ecoenv.2016.10.006.

Zhao, P., Zhang, X., Sun, C., Wu, J. & Wu, Y. (2017) Experimental study of conservative solute transport in heterogeneous aquifers. *Environmental Earth Sciences*. [Online], 76, 421. doi:10.1007/s12665-017-6734-2.

Zuccato, E., Castiglioni, S., Bagnati, R., Melis, M. & Fanelli, R. (2010) Source, occurrence and fate of antibiotics in the Italian aquatic environment. *Journal of Hazardous Materials*. [Online], 179, 1042–1048. doi:10.1016/j.jhazmat.2010.03.110.

Comparison and validation of different methods of groundwater vulnerability assessment for different groundwater systems

Chapter 9

Groundwater vulnerability and risk assessment in Kaduna metropolis, northwest Nigeria

M.S. Ahmed, A.I. Tanko, M.M. Badamasi & A. Abdulhamid

1.0 Introduction

Towns and cities in developing countries, such as Nigeria, lack adequate portable water supplies, as such groundwater extraction for both domestic and other uses is essential. In Kaduna metropolis for example, efforts to meet the domestic demand of water supply from the piped water system by successive governments proved unsuccessful. Despite the growing importance of groundwater in supporting human livelihood and ecological balance, threats to its quality are being heightened in recent times by anthropogenic activities. The most significant sources of contaminants are those related to urbanisation and agriculture. Municipal dumpsites, landfills, underground storage tanks, urban runoff, septic systems, agricultural chemicals (Zaporezec, 2002) and industrial activities, among others, pose serious concern for groundwater quality.

Groundwater, unlike surface water, has a natural shield against contamination, it is thus, not easily contaminated. However, once the quality is impaired, it proves difficult or nearly impossible to be remediated. A proactive measure to prevent groundwater contamination is, therefore, the basis of sustainable groundwater quality management. Effective groundwater protection aims at preventing contamination, based on sound information about groundwater problems that exist and those that may develop in the future (Gibrich and Zaporezec, 1994). Some natural physical characteristics of an area can provide significant protection to the groundwater. The degree of protection varies spatially and thus, determines the vulnerability of groundwater to contamination. However, actual contamination of groundwater depends not only on how vulnerable the area is to contamination, but to a large extent on the presence of contaminant sources and the interactions between the two variables. Contamination risk implies the possibility of groundwater contamination due to presence of contaminant (hazard) sources. This according to Daly *et al.* (2004), depends on the hazard posed by a potential polluting activity, the intrinsic vulnerability of groundwater to contamination, and the potential consequences of a contamination event. Assessing groundwater contamination risk involves assessing and mapping groundwater vulnerability and includes land use as a risk indicator. Both simple and relatively sophisticated approaches have been used by several researchers for this assessment. While some researchers such as Alhanbali and Kondoh (2008); Jasem and Alraggad (2010) and Al-Adamat *et al.* (2003) simply included a land use rating to the vulnerability assessment, others such as Ducci (2009); Fadlelmawla *et al.* (2011) and Wang *et al.* (2012) used more standardised approaches. The most advanced model of groundwater contamination risk assessment is perhaps the EU COST Action 620 otherwise known as the Pan European approach (Zwahlen, 2004) which provides a step by step guide on assessment. The development of hazard and risk assessment method and mapping by Working Group 3 proves its applicability elsewhere as a valuable tool in groundwater protection. According to this approach, assessing groundwater contamination risk involves three separate tasks, vulnerability assessment and mapping, hazard inventory and mapping and risk assessment. Apart from its application in the COST 620 test sites such as Seirra de Libar in Southern Spain (Andreo *et al.*, 2006), Swabian Alb, Germany (Goldscheider, 2005) among other places, other researchers such as Mimi and Assi (2009); Werz (2006) and Nguyet and Goldscheider (2006) achieved better result in the application of this approach in different places.

While many cases of groundwater contamination have been reported in Kaduna metropolis, the potential contaminant sources and the risk they pose to groundwater in the area are not well known and documented (Ahmed, 2016). Since groundwater serves as the best potable source of water in the area, a viable means of protecting this resources via identification and mapping the risk levels is needed. This research documented the hazards and assessed their risk to groundwater contamination. DRASTIC and GOD models as well as the Pan European approach to risk intensity mapping (Zwahlen, 2004) were used in the assessment.

2.0 Methods

2.1 The study area

Kaduna metropolis is one of the major cities in northern Nigeria. Located on latitude 10^0 18′ 40″- 10′ 40′ 48″ north of the equator and longitude 7^0 11′ 6″ – 7^0 36′18″ east of the Greenwich meridian (Figure 9.1). The metropolis lies at an altitude of about 643m above sea level. It comprises the whole of Kaduna North and South, and parts of Igabi and Chikun local government areas within Kaduna State. The area stretches from Katabu in the north to Sabon Gayan in the south, to Buruku and Kujama in the west and east respectively representing the area defined as Kaduna city region by Max Lock and Partners (2008). The climate of Kaduna is tropical wet and dry coded as Aw by Koppen with rainfall of about 1200mm annually which typically last between 5 to 6 months (April to September). The rainy season is preceded by a short hot dry spell with mean monthly temperatures of between 35°C and 40°C (Mallo, 2001). Temperature is generally hot throughout the year with the exception of slight period of cold and dry weather (November to February).

The area lies largely within the lower Kaduna catchment with the geology predominantly metamorphic rocks of the Nigerian basement complex rock composing mostly migmatite-gneiss complex and meta-sedimentary series. Groundwater occurrence is mostly in the weathered/ fractured basement complex and river alluvium (Eduvie, 2003). The weathered metamorphic and magmatic rocks produce weathered products known as regolith, saprolite or alterite comprising a mixture of sands and clays of varying thickness overlaying the altered or fractured parent rock (Eduvie, 2003). Typical layers in the saprolite profile are lateritic soil, over clay alterite layer, granuler sandy zone and bedrock at the base (Jones in Eduvie, 2003). The relief is mostly undulating plains with isolated high plains in some parts. The soils are red brown to red yellow ferruginous soils. The vegetation of the area is northern guinea savannah with predominant grassland and scattered trees. Nearly all Nigerian ethnic groups can be found in the metropolis. Urban agglomeration is put at 1,422,000 by UN estimate for 2007 (Max Lock Consultancy Nig. Ltd. and Partners,

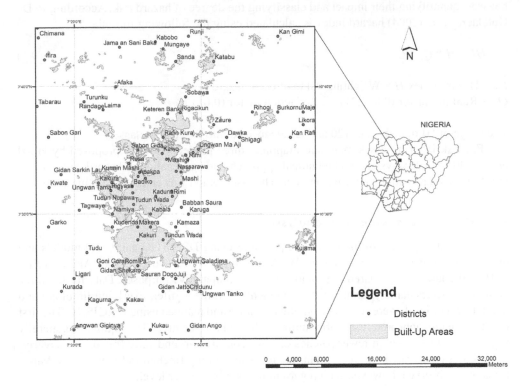

Figure 9.1 Kaduna metropolis.

2008). The population of Kaduna metropolis was estimated to be about 2,057, 000 by 2015 with females of reproductive age (15–49) comprising about 48% and males comprising the remaining 52% (Salim in Saleh, 2015). The metropolis is a home to many industries such as petroleum refining, textile, Peugeot assembly plant, brewery, flour mills, cable, iron and steel, fertiliser, aluminium industries and others. Farming activities are mostly concentrated along the flood plain of River Kaduna and its tributaries. Other forms of farming, as noted by Saleh (2015), include domestic gardens, market gardens, peri-urban/open space farming, allotment gardens, school gardens, rail and road verge farms, recreational gardens, undeveloped plot cultivation and vacant farm plots.

2.2 DRASTIC index, GOD and EU COST Action 620 approach

The DRASTIC index model (Aller *et al*., 1987) is a parametric model that evaluates the significance of depth to water level, net recharge, aquifer media, soil, topography, impact of the vadose zone and hydraulic conductivity. GOD (Foster, 1987) considers the significance of groundwater confinement, overlying lithology of the aquifer and depth to water level. Details about the models can be found in Aller *et al*. (1987) and Foster (1987) respectively. The EU approach (Zwahlen, 2004) to hazard and risk assessment provides a framework for effective risk assessment. The assessment normally starts with the identification of potential hazards, quantifying their impact and classifying the degree of hazard risk. According to De Ketelaere *et al.* (2004) hazard index is calculated using the following formula:

$$HI = H * Q_n * R_f \tag{9.1}$$

HI = Hazard index *H* = Weighting value of each hazard
Q_n = Ranking factor (0.8–1.2) R_f = Reduction factor (0.0–1.0)

The range of HI runs from 0–120 and is categorized into five or six classes.

For risk intensity assessment and mapping, three procedures were proposed by Hotzl *et al.* (2004), matrix system, mathematical approach and aggregation of hazard, and vulnerability base maps within GIS environment. The last method was employed in this study.

2.3　Data type, sources and analysis

Two vulnerability maps of the area were prepared using DRASTIC and GOD models and used in the vulnerability assessment. While the former was sourced from Ahmed *et al.* (2017), the later was prepared to provide data for comparative purposes. Data for the three GOD parameters namely, groundwater confinement in the aquifer, overlying lithology and depth to water level, were sourced from the literature and mapped using ArcGIS 10. The first two parameters appeared reasonably uniform throughout the area. The aquifer is generally unconfined (rated 1) with fewer patches of semi-confined groundwater (rated 0.4). Overlying lithology was rated 0.6 throughout due to its uniformity (metamorphic/igneous). Variations were observed, however, with regard to the depth to water level.

Using the contaminants source inventory list provided by Cost Action 620, different land uses in the study area were identified and their weight (H) assigned. Information about some existing contaminant sources such as dumpsite, hazardous dumpsite, polluted rivers, wastewater irrigation, and a list of industries were obtained from Kaduna State Environmental Protection Agency (KEPA) and the state's ministry of Agriculture. Petrol stations and pipeline data were secured from the Department of Petroleum Resources (DPR) and Pipeline and Products Marketing Company (PPMC) respectively. List of automobile garages was secured from the Nigeria Automobile Technician Association (NATA), Kaduna state chapter. Hazard locations were visited for quantification, ranking as well as determining their coordinates. Other information were extracted from satellite imagery. The identified hazard sites were visited to determine their coordinates. They were mapped and field checked to assess their properties with respect to the quantity of contaminating substances and any likely reduction factor. Access to most of the industries however, was difficult, as such they were rated without considering a reduction factor but ranked according to their size. The ranking factor (Qn) and the reduction factor (Rf) were estimated on the bases of relative size and the technical conditions of the hazard.

A vector data model with points, lines and polygons of ArcGIS 10.0 was used in the hazard and risk mapping. A database of each contaminant source (hazard) was created taking into consideration its spatial properties. Hazard index calculation was done by entering the required coefficients (H, Qn and Rf) in the form of attributes. It was calculated with the "calculator" tool in GIS and stored in a separate column. With this, a classified map depicting the hazards according to their HI was produced. A map of hazard distribution (unclassified hazard map) was also prepared to show their spatial distribution. An overlay of the classified hazard map and vulnerability map produced the risk intensity (RI) map. Groundwater samples were collected from the refinery area for Gas Chromatography-Mass Spectrometry (GC-MS) analysis to ascertain the presence or otherwise of benzene, toluene, ethylbenzene, xylene (BTEX).

3.0 Result and discussion

3.1 Groundwater vulnerability in Kaduna metropolis

A map of groundwater vulnerability of the area using the GOD model was produced using an overlay with ArcGIS. About 80% of the area has moderate vulnerability to contamination. High vulnerability (0.54) covers about 18% while the remaining land falls within the low vulnerability (0.26) area (Figure 9.2). The DRASTIC index values (77–131) in the area also show low to moderate vulnerability (Figure 9.3).

3.2 Hazard assessment

Sources of groundwater contaminants (hazards) in Kaduna metropolis are diverse. Information about the accessible hazards covered by this research are summarised in Table 9.1.

The entire residential area falls within the category of urbanisation without a sewage system, and has a hazard weighting value of 70. A ranking factor (Qn) of 0.8 was used to adjust the values in the low density residential areas to 56. Domestic wastewater is normally

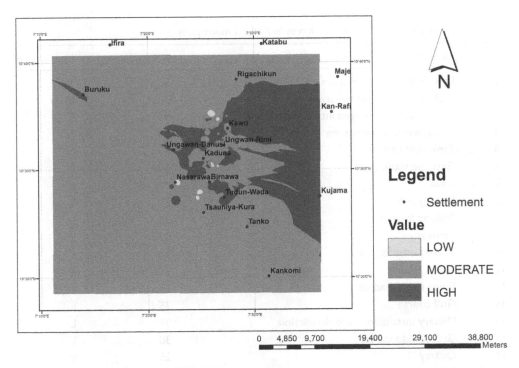

Figure 9.2 GOD vulnerability map of Kaduna metropolis.

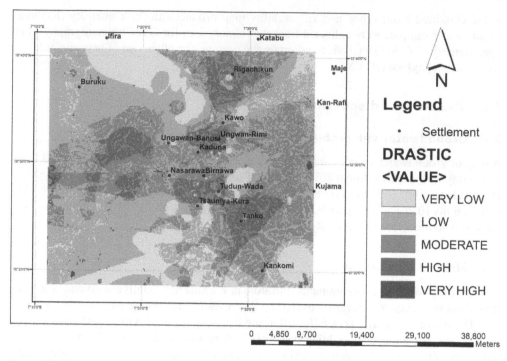

Figure 9.3 DRASTIC index Map of Kaduna Metropolis (Ahmed *et al.*, 2017).

Table 9.1 Groundwater contaminant sources in Kaduna metropolis

S/N	Hazard	Hazard Index	Hazard Level
1	Low density residential areas	56	M
2	High density residential areas	70	M
3	Wastewater discharged into surface water courses	45	L
4	Dumpsites/Hazardous waste site	90	H
5	Tank Yard (major)	50	M
6	Gasoline station	60	M
7	Road	40	L
8	Railway line	30	L
9	Railway Station	35	L
10	Runway	35	L
11	Tourist urbanisation	30	L
12	Camp ground	30	L
13	Open sport stadium	25	L
14	Golf course	35	L
15	Military installations and dereliction	35	L
16	Gravel and sand pit	30	L
17	Quarry	25	L
18	Oil Pipeline	55	M

S/N	Hazard	Hazard Index	Hazard Level
19	Oil refinery	85	H
20	Wastewater irrigation	60	M
21	Chemical industries	65	M
22	Metal processing industries	50	M
23	Textile industries	65	M
24	Food and beverages industries	45	L
25	Iron, steel and aluminium industries	40	L
26	Electroplating industry	55	M
27	Rubber/plastic industries	40	L
28	Glass and porcelain industries	40	L
29	Foam production industries	45	L
31	Oil blending industries (surface tank usage)	50	M
32	Automobile (mechanic) Garage	50	M

discharged into the surface water bodies without any prior treatment. This is corroborated by Ekiye and Zejiao (2010); Akpan *et al.* (2008) and Mahre *et al.* (2007) such that the River Kaduna is the sink for pollutants in the Kaduna metropolis. The same can be said about the industrial effluents which mostly exceeded the regulatory standards at the discharge points (Al-Amin, 2013; Lekwot *et al.*, 2012). Wastewater irrigation was observed along the River Romi, behind Kakuri and Kudenda industrial areas and elsewhere. Small dumpsites within the residential areas were excluded partly due to the absence of a database, as well as the ongoing mobile refuse collection programme in the area. While the two tank yards at Buruku and Unguwan Mu'azu were weighted 50 accordingly, a ranking factor of 1.2 was used to adjust the weighting of the Maraban Jos yard to 60 in order to reflect its size and the volume of tankers involved.

About 342 gasoline (filling) stations of varying sizes were found in Kaduna metropolis. Automobile (mechanic) garages, though not considered by the model, were included in the hazard assessment in this study. The garages which are normally unpaved, were noticed to be heavily polluted by used motor oil among other substances found in the repair and painting shops. Only those garages found in the register of NATA were considered in this assessment. A hazard weighting value of 50 was assigned to this category, however, the three "mechanic villages" at Oria Kpata, Kurmin Mashi and Mando (Units 1, 2 and 13A,B and C respectively) were ranked higher (Qn = 1.2) to reflect their relative size and mixture of activities. A reduction factor (Rf = 0.9) was applied to mechanic garage at Kawo park because of its well paved nature of the surface.

About 50 industries of different types and sizes were surveyed in this study. They are particularly located at the Kakuri and Kudenda industrial areas, others were found at the Hayin Banki and Mando light industrial areas. The industries include food and beverages, chemical, iron and steel, metal processing, electroplating, textile, rubber/plastic, foam production, glass and porcelain, oil refining, storage and blending industries. All the industries were weighted according to their original weighting values (De Ketelaere *et al.*, 2004) as information regarding any reduction factor was scarce and did not warrant value judgement. Considering the volume of liquid handled in the beverages industries, their ranking factor

(Qn) was adjusted higher (1.2) than the other food industries, the Qn was however, adjusted lower (0.8) for smaller cooking oil production factories. Two types of industries not included in the COST Action 620 list, the foam production as well as glass and porcelain industries were weighted 45 and 40 respectively. Kaduna refinery was weighted 85, while the pipeline that supplies crude to the refinery and the product lines to Suleja, Jos, Kano- Gusau depots were weighted 55. Textile industries were rated 65, this was deduced from the original rating for chemical factories.

Other hazards surveyed during the study include the camp ground at Mando Hajj camp, tourist urbanisation centre at Gamji Gate, Polo ground at Maraban Jos, Golf course at Ungu-wan Rimi, sport stadia (Ranchers Bees, Township and Murtala Square) and several military installations. Although sand excavation and quarrying in the area are considered illegal by KEPA, the former is carried out along the accessible areas of River Kaduna, while the latter is undertaken where there is rock exposure such as at Malali, Unguwan Rimi, along the Jos road and at Kujama. Roads, a railway line and airport runway were weighted accordingly with a buffer of 10 m since most of the contamination is likely to emanate from runoff from the traffic areas.

Two hazard maps, unclassified and classified were produced. The unclassified hazard map (Figure 9.4) shows the distribution of the identified hazards in the study area while the classified hazard map (Figure 9.5), shows the various classes of the hazards based on the hazard assessment and mapping procedure of the EU Cost Action 620. From the classified

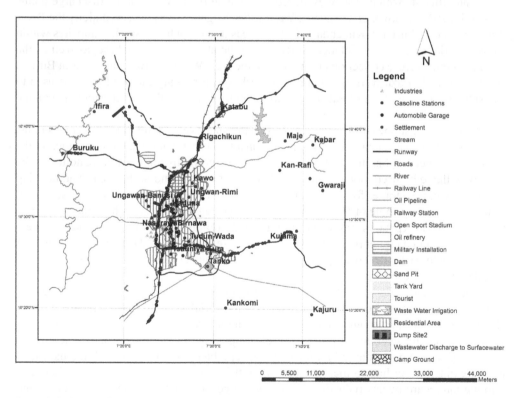

Figure 9.4 Unclassified hazard map of Kaduna metropolis

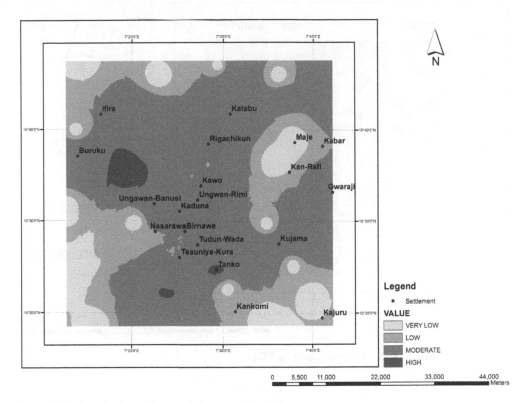

Figure 9.5 Classified hazard map of Kaduna metropolis

hazard map (Figure 9.5), it is evident that only the areas of Kaduna refinery and the two dumpsites have high hazard levels and, therefore, appear orange in colour. The majority of the built-up areas have moderate hazards (yellow colour) representing the highest hazard in the area (urbanisation without sewer). Low hazard (green colour) occur within the built-up areas representing the military installations, as well as the cultivated areas. Most of the cultivated and uncultivated forested areas appeared blue representing no or very little hazard.

3.3   Risk mapping

This research, takes into account the likelihood of an impact (hazard analysis), and intensity of a potential impact (risk intensity). The sensitivity of groundwater with respect to the impact (risk sensitivity) which requires information on groundwater value was not included. Groundwater is a highly valuable resources in the whole study area; any impairment as regards the quality of groundwater in the study area will have a negative impact on both the social and economic life of the inhabitants and could create serious ecological problem. The vulnerability and classified hazard maps were overlaid within a GIS environment to produce the risk maps based on GOD (Figure 9.6) and DRASTIC (Figure 9.7) models. High groundwater contamination risk can be seen in the eastern and central parts of the study area as a reflection of the high vulnerability and hazards of the areas. Moderate risk class, which

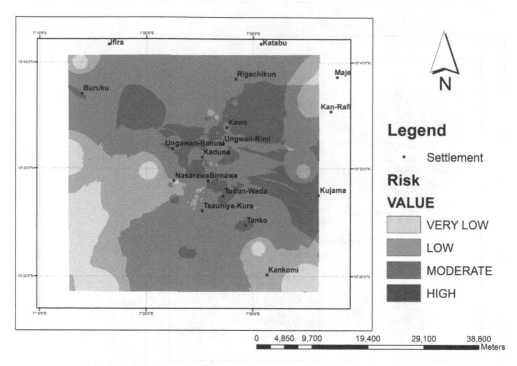

Figure 9.6 Risk map of Kaduna metropolis using GOD

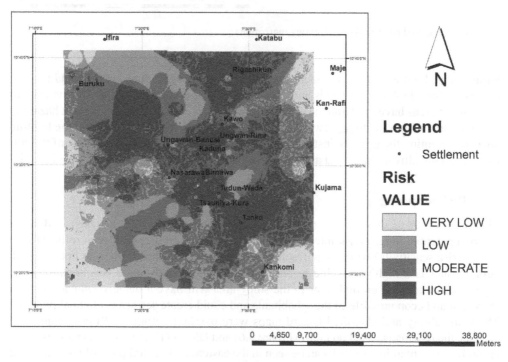

Figure 9.7 Risk map of Kaduna metropolis using DRASTIC

dominates the whole study area, conforms with the overall moderate status of both vulnerability and hazard maps. Low and very low contamination risk reflects the hazard distribution of the study area.

3.4   Hydrocarbon contaminants in groundwater

The hazard and risk maps indicated three outstanding high risk areas: the two dumpsites and the refinery. GC-MS analysis of the groundwater samples was conducted to ascertain the presence or otherwise of BTEX which is believed to be diagnosis good indicator of oil spills (Fetter in Kralik *et al.*, 2004). No individual compound of BTEX was found, benzene derivative compound such as benzaldehyde, 2-hydroxy-4-methyl- ($C_8H_8O_2$), benzenemethanol, 4-nitro- ($C_7H_7NO_3$), Benzene, [(2,2-dimethylcyclopropyl)methyl]- ($C_{12}H_{16}$) were however, found in one of the samples. The three other samples contain acids, esters, alcohols and aldehydes. This is in contrast with Oil Mop (1982) whose findings revealed that, the refinery operation has no significant impact on groundwater.

4.0   Conclusion

The groundwater resources in most parts of Kaduna metropolis is low to moderately vulnerable to contamination. High risk of contamination was however observed in many areas due to the presence of significant pollution hazards. The high risk zones conform with the industrial and residential areas. This is due to the absence of sewer system and poor waste management in the metropolis couple with the moderate vulnerability. The presence of benzene derivative compounds in groundwater around the refinery confirms that the location is at high risk of contamination. A detailed study of organic contaminants in the area could better define the extent of contamination. Future land use with high groundwater potential hazards are recommended in the low vulnerability areas. Where an existing land use with a higher contamination hazard is already located on a highly vulnerable area, as in the case of the refinery and the dumpsites, more sophisticated control measures should be put in place. This can be achieved through proper monitoring and enforcement of environmental regulations. Environmentally friendly, engineered dumpsites are urgently required in the area to replace the two open dumpsites of Kakau and Mando.

References

Ahmed, M.S. (2016) *Assessment of Groundwater Vulnerability and Risk of Hydrocarbon Contamination in Kaduna Metropolis, Nigeria.* Unpublished PhD Thesis, Department of Geography, Bayero University, Kano, Nigeria.

Ahmed, M.S., Tanko, A.I., Eduvie, M.O. & Ahmed, M. (2017) Assessment of groundwater vulnerability in Kaduna metropolis, northwest Nigeria. *Journal of Geoscience and Environment Protection*, 5, 97–115.

Akpan, U.G., Afolabi, E.A. & Okemini, K. (2008) Modeling and simulation of the effect of effluent from Kaduna refinery and petrochemical company on River Kaduna. *Assumption University Journal of Technology AU J.T.*, 12(2), 98–106.

Al-adamat, R.A.N., Foster, I.D.L. & Baban, S.M.J. (2003) Groundwater vulnerability and risk mapping for the basaltic aquifer of the Azraq Basin of Jordan using GIS: Remote sensing and DRASTIC. *Applied Geography*, 23, 303–324, Elsevier.

Al-Amin, M.A. (2013) Energy production and environmental concerns in Nigeria: The case of Kaduna petroleum refinery on its host communities. *Journal of Energy Technologies and Policies*, 3(10), 69–77.

Alhanbali, A. & Kondoh, A. (2008) Groundwater vulnerability assessment and evaluation of human activity impact (HAI) within the dead sea groundwater basin, Jordan. *Hydrogeology Journal*, 16, 499–510, Springer-Verlag.

Aller, A., Lehr, J.H., Petty, R. & Bennet, T. (1987) *DRASTIC: A standardized system to evaluate groundwater pollution potential using hydrogeologic settings*. National Water Well Association, Wethington, Ohio.

Andreo, B., Goldscheider, N., Vadillo, I., Vias, J.M., Neukum, C., Sinreich, M., Jime´nez, P., Brechen-macher, J., Carrasco, F., Hotzl, H., Perles, M.J. & Zwahlen, F. (2006) Karst groundwater protection: First application of a Pan-European approach to vulnerability, hazard and risk mapping in the Sierra de Libar (Southern Spain). *Science of the Total Environment*, 357, 54–73, Elsevier.

Daly, D., Hotzl, H. & De Ketalaere, D. (2004) Risk definition. In: Zwahlen, F. (ed.) *Vulnerability and Risk Mapping for the Protection of Carbonate (Karst) Aquifers*, final report (COST action 620), Report EUR 20912. European Commission, Directorate XII Science, Research and Development, Brussels. pp. 106–107.

De Ketelaere, D., Hotzl, H., Neukum, C., Civita, M. & Sappa, G. (2004) Hazard analysis and mapping. In: Zwahlen, F. (ed.) *Vulnerability and Risk Mapping for the Protection of Carbonate (Karst) Aquifers*, EUR 20912. European Commission, Directorate-General XII Science, Research and Development, Brussels. pp. 86–105.

Ducci, D. (2009) GIS techniques for mapping groundwater contamination risk. *Natural Hazards*, 20, 279–294.

Eduvie, M.O. (2003) *Exploration, Evaluation and Development of Groundwater in Southern Kaduna State*. Unpublished PhD Thesis, Department of Geology, Ahmadu Bello University, Zaria.

Ekiye, E. & Zejiao, L. (2010) Water quality monitoring in Nigeria: Case studies of Nigeria's industrial cities. *Journal of American Science*, 6(4), 22–28.

Fadlelmawla, A.A., Fayad, M., El-Gamily, H., Rashid, T., Mukhopadhyay, A. & Kotwicki, V. (2011) A land surface zoning approach based on three-component risk criteria for groundwater quality protection. *Water Resource Management*, 25, 1677–1697.

Foster, S.S.D. (1987) Fundamental concepts in aquifer vulnerability, pollution risk and protection strategy. In: Duijvenbooden, W. & Waegeningh, H.G. (eds.) *Vulnerability of soil and groundwater to pollutants*. TNO Committee on Hydrological Research, The Hague, Proceedings and Information 38. pp. 69–86.

Gibrich, W.H. & Zaporezec, A. (1994) Introduction. In: Vrba, J. & Zaporezec, A. (eds.) *Guidebook on Mapping Groundwater Vulnerability*, Volume 16. International Contribution to Hydrogeology, Hannover. pp. 1–2.

Goldscheider, N. (2005) Karst groundwater vulnerability mapping: Application of a new method in the Swabian Alb, Germany. *Hydrogeology Journal*, 13, 555–564, Springer.

Hotzl, H., Delporte, C., Liesch, T., Malik, P., Neukum, C. & Svasta, J. (2004) Risk mapping. In Zwahlen, F. (ed.) *Vulnerability and Risk Mapping for the Protection of Carbonate (Karst) Aquifers*, final report (COST action 620), Report EUR 20912. European Commission, Directorate XII Science, Research and Development, Brussels. pp. 113–120.

Jasem, A.H. & Alraggad, M. (2010) Assessing groundwater vulnerability in Azraq basin area by a modified DRASTIC index. *Journal of Water Resource and Protection*, 2, 944–951, Scientific Research.

Kralik, M., Kranjc, A. & Meus, P. (2004) Organic contaminants. In: Zwahlen, F. (ed.) *Vulnerability and Risk Mapping for the Protection of Carbonate (Karst) Aquifers*, final report (COST action 620). European Commission, Directorate XII Science, Research and Development, Report EUR 20912, Brussels.

Lekwot, V.E., Caleb, A.I. & Ndahi, A.K. (2012) Effects of effluent discharge of Kaduna refinery on the water quality of River Romi. *Journal of Research in Environmental Science and Toxicology*, 1(3), 41–46.

Mahre, M.Y., Akan, J.C., Moses, E.A. & Ogugbuaje, V.O. (2007) Pollution indicators in River Kaduna, Kaduna state, Nigeria. *Trends in Applied Sciences Research*, 2(4), 304–311.

Mallo, I.Y. (2001) Morphometric characteristics of Barnawa River Catchment in Kaduna Metropolis, Northern Nigeria. *Journal of Environmental Sciences*, 4(1), 22–28, Faculty of Environmental Sciences, University of Jos, Nigeria.

Max Lock Consultancy Nig. Ltd. and Partners (2008) *Kaduna Master Plan Review Interim Report*, Kaduna.

Mimi, Z.A. & Assi, A. (2009) Intrinsic vulnerability, hazard and risk mapping for Karst aquifers: A case study. *Journal of Hydrolology*, 364, 298–310.

Nguyet, V.T.M. & Goldscheider, N. (2006) Simplified methodology for mapping groundwater vulnerability and contamination risk, and its first application in a tropical Karst area, Vietnam. *Hydrogeology Journal*, 14, 1666–1675.

Oil Mop (1982) *Report on Environmental Study of the NNPC Locations and Recommendations for Anti-Pollution Contingency Plan*. University College London Press, London.

Saleh, Y. (2015) *Kaduna: Physical and Human Environment*. Shanono Printers and Publishers, Kaduna.

Wang, J., He, J. & Chen, H. (2012) Assessment of groundwater contamination risk using hazard quantification, a modified DRASTIC model and groundwater value, Beijing Plain, China. *Science of the Total Environment*, 432, 216–226, Elsevier.

Werz, H. (2006) The use of remote sensing imagery for groundwater risk intensity mapping in the Wadi Shueib, Jordan. Unpublished PhD Thesis, Department of Applied Geology, University of Karlsruhe.

Zaporezec, A. (2002) *Groundwater Contamination Inventory: A Methodological Guide*. IHP-VI Series on Groundwater No. 2, UNESCO.

Zwahlen, F. (2004) *Vulnerability and Risk Mapping for the Protection of Carbonate (Karst) Aquifers*, final report (COST action 620). European Commission, Directorate XII Science, Research and Development, Report EUR 20912, Brussels. p. 297.

Chapter 10

Evaluation of the comparison of four groundwater vulnerability methodologies

A case study of Dahomey Basin shallow aquifers, Nigeria

S.A. Oke & D. Vermeulen

I   Introduction

Groundwater vulnerability assessments are means of ensuring effective protection of groundwater resources and sources. Vulnerability assessments are done using vulnerability methodologies. Vulnerability methods take into accounts parameters that influence and contribute to groundwater risk factors. These parameters are pollutants, climatic conditions, anthropogenic activities and the general intrinsic properties of an aquifer. An effective vulnerability method must simplify hydrogeological parameters into zoning which can be compared with the observation results (Faybishenko *et al.*, 2014).

Several methods have been developed to assess groundwater vulnerability (Gogu and Dassargues, 2000; Goldscheider, 2002; Liggett and Talwar, 2009; Wachniew *et al*, 2016). Some of these methods are more effective than others especially when assessing heterogeneous aquifer systems. The European concept is based on the Origin-Pathway-Target model; this is contained in the Cost Action 620 report (Daly *et al.*, 2002). Intrinsic vulnerability concepts distinguish degrees of vulnerability at regional scales where different lithologies exist (Vias *et al.*, 2005; Wachniew *et al.*, 2016). Common intrinsic vulnerability methods are subjective (overlay or index) methods. Travel time concept also referred to as transit time, seepage time, turnover time and residence times is a physically based model that uses time-based methodologies to assess vulnerability of groundwater sources and resources.

Due to the diverse background groundwater vulnerability concepts are based, it becomes challenging chosen a suitable methodology that will effectively assess shallow aquifer vulnerability. Therefore, this paper aims to examine four selected vulnerability methods effectiveness, assessment parameters, inter-relations and classification through application to a shallow sedimentary aquifer. This will be done through evaluating the methods comparisons.

I.I   *Vulnerability methodologies*

The four vulnerability methodologies applied to evaluate the shallow aquifers of the Dahomey Basin of southwestern Nigeria are the DRASTIC, PI, AVI and RTt method. The methods are based on the European approach to vulnerability mapping (COST Action 620, 2003) which is based on the concept of origin-pathway-target models. Others include the index-based concept and travel time concept of vulnerability assessments. The travel time concepts mirrored a steady state conditions which are commonly used in the physically based vulnerability methods (Oke and Fourie, 2017).

2   Materials and method

2.1   DRASTIC methodology

The DRASTIC method was developed by Aller *et al.* (1987). It is widely used in intrinsic vulnerability assessment of aquifers. DRASTIC evaluate seven intrinsic parameters which include the depth to water, groundwater recharge, aquifer characteristics, soil type, topographic condition, vadose zone and hydraulic conductivity. DRASTIC employs the rating and weighting system based on the Delphi technique (Aller *et al.*, 1987). DRASTIC is calculated as follows:

$$DRASTIC\ Index = D_R D_W + R_R R_W + A_R A_W + S_R S_W + T_R T_W + I_R I_W + C_R C \tag{10.1}$$

where:
D, R, A, S, T, I, and C are the seven parameters of the model, subscripts R and W are the corresponding ratings and weights, respectively.

The seven hydrogeological parameters rating ranged from 1–10. This implies the higher the DRASTIC rating and index, the greater the risk of groundwater to contamination. DRASTIC requires that the knowledge of system to be assessed. Its other drawbacks include that DRASTIC is not based on a clear conceptual model such as the origin-pathway-target model of European Vulnerability Model (COST Action 620, 2003) and DRASTIC over emphasises slope.

2.2   AVI method

AVI method was developed by Van Stempvoort *et al.* (1993) and approved by the Canadian Prairie Province Water Board. AVI strength lies on vadose zone characterisations. Aquifer vulnerability is computed based on hydraulic resistance (c) which is a ratio between the thickness of each sedimentary bed unit above an aquifer and the hydraulic conductivity of each bed. Hydraulic resistance is calculated as

$$c = \sum_{i=1}^{n} \frac{d_i}{K_i} \tag{10.2}$$

where:
n = number of sedimentary units above the aquifer
d_i = thickness of the vadose zone
K_i = hydraulic conductivity of each protective layer
K = unit of length/time (m/s or m/d)
c = travel time with dimension in seconds

2.3   RTt method

RTt method was based on a simplified concept of advective travel time of rainfall to shallow groundwater. It was developed by Oke *et al.* (2016). RTt method conceptual aim was to consider important parameters that influenced the recharge of groundwater by rainfall through the unsaturated zones. Travel time was calculated based on modified Darcy's equations as follows:

$$Tt = \sum_{i=1}^{n} \frac{D*S}{\left(\dfrac{K_{sat}}{\theta}\right)}$$

(10.3)

where:
Tt = travel time
D = depth from ground surface to aquifer (m)
K_{sat} = hydraulic conductivity at saturation of successive layers (m/s)
Θ = effective porosity of the medium
S = slope (elevation head difference: dh/dl) in meters
N = numbers of layers between ground surface and the top of the aquifers

RTt intrinsic vulnerability is calculated based on the weight and rating system of travel time and rainfall quantification as follows:

$$RTt = R_R R_w + Tt_R Tt_W$$

(10.4)

where:
R_R = rainfall rating = 10
R_w = rainfall weight
Tt_R = travel time rating = 10
Tt_W = travel time weight

2.4 PI method

The PI method developed by Goldscheider *et al.* (2000) was designed for vulnerability assessment of both karst and non-karst aquifer systems. The PI method followed the concept of the origin – pathway – target model of the European groundwater vulnerability assessment (COST Action 620, 2003). P stands for protective cover of the strata above the aquifer while I described the infiltration conditions, which is the degree to which the protective cover is bypassed as a result of both lateral and subsurface flow. The final protection factor π is calculated based on the product of P and I. It is subdivided into five classes. A protective factor of $\pi \leq 1$ indicates a very low degree of protection and an extreme vulnerability to contamination; $\pi = 5$ indicates a high degree of protection and a very low vulnerability.

2.5 Application to Dahomey Basin

The evaluation of the four methods was done on the regional sedimentary basin of southwestern Nigeria. One of the fundamental concept of groundwater vulnerability is that some areas are prone to contamination than others (Oke and Fourie, 2017) and intrinsic vulnerability assessment which these comparisons is based, is more of in-situ geological system examination of the ease of contaminant infiltrating and percolating a geological system to contaminate shallow groundwater (Oke *et al.*, 2018). Therefore, the application of the methods to

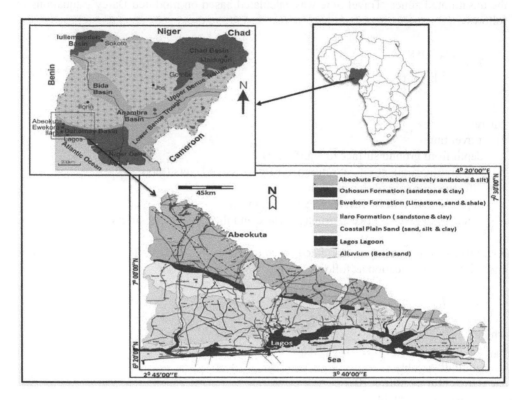

Figure 10.1 Geological map of the Dahomey Basin of Nigeria

the Dahomey Basin is to predict the area of the basin most prone to infiltrating contaminant and to ascertain the most effective method for assessing shallow aquifers. The geohydrological condition of the basin includes 3 hydrostratigraphic units as stated by Onwuka (1990); these units are the upper aquifer (alluvium and coastal plain sands), middle aquifer (Ilaro and Ewekoro formations) and lower aquifer (Abeokuta Formation) (Figure 10.1).

The basin hydraulic head gets shallower towards the sea. The nature of the geological formations plays a significance role with the depth to water level. Abstraction rate shows a yield ranging from 21.5–49.2 m³/hr and 10–58 m3/hr (Offodile, 2014) for the Alluvium and Ewekoro Formation. The Abeokuta Formation yield record 1.6 m³/hr were the formation overlies the Basement rocks to 10–36.3 m³/hr were it consists of thick deposited sediments.

3   Results

3.1   Comparison of the vulnerability methods

The four vulnerability maps show some similarities as well as differences. A comparison of the parameters considered in the vulnerability methods is presented in Table 10.1. The AVI method has the least considered parameters, while the PI method has the most considered parameters. Only the RTt and AVI considered the travel time. While all the methods were developed for resource vulnerability assessments only the PI methods was developed for source vulnerability assessment. Source vulnerability involves assessments of groundwater sources in the well or spring including karst networks and resources assessment target groundwater resources and the overlying layers above the aquifer (Daly *et al.*, 2002). Depth of unsaturated zone, subsoil thickness and soil permeability were as well considered by all the methods.

The PI, RTt and DRASTIC methods considered topsoil thickness, rock types, recharge and slope gradient. Only the RTt method considered the effective porosity of the rock materials. All the methods considered the depth to the water table, subsoil permeability and subsoil thickness. Land use and flow concentrations were considered in the PI method. Determination of flow path is important to PI method due to PI objectives of assessing karst topography (Goldscheider, 2002).

The RTt, PI and DRASTIC method were based on the Point Count System (PCSM) models. The PCSM model determine vulnerability by rating parameters assumed to be significant to vulnerability and assigning values to these parameters. This is different to the European concept to vulnerability further used by the PI and RTt methods. The European concept though developed for karst systems (COST Action 620, 2003) is applicable to non-karst areas. The concept suggested vulnerability should be based on the origin-pathway-target model of environmental management (Daly *et al.*, 2002).

Table 10.1 Parameters considered under the four vulnerability method in this research

Parameters	DRASTIC	PI	RTt	AVI
Topsoil thickness	X	X	X	O
Topsoil texture	O	X	X	O
Topsoil structure	O	X	O	O
Subsoil permeability	X	X	X	X
Subsoil thickness	X	X	X	X
Depth of the unsaturated zone	X	X	X	X
Fracturing	O	X	X	O
Epikarst development/ geomorphological features	O	X	O	O
Travel time estimation	O	O	X	X
Confined condition	X	X	O	O
Concentration of flows	O	X	O	O
Slope gradient	X	X	X	O
Land use/vegetation cover	O	X	O	O
Recharge	X	X	X	O
Hydraulic characteristics of aquifer	X	O	X	O
Effective porosity	O	O	X	O
Resources vulnerability	X	X	X	X
Source vulnerability	O	O	O	O

3.2 Comparison of the vulnerability maps and class

The RTt vulnerability assessment map and class of the Dahomey Basin shows 18% as very high vulnerability, 7% as high vulnerability, 64% as moderate vulnerability and 11% as low vulnerability (Figure 10.2). The very high vulnerabilities are areas within the lowland and swamp environments. The high area is characterised by rivers, interconnected streams and low depth to water range of 0–5 m. Almost two-third of the basin was classified as moderate vulnerability due to the low slopes, soil and geological formations (bedded sandstone) which reduces infiltration even though rainfall was general high between 1190 mm – 1618 mm (Oke *et al.*, 2016).

According to the PI map and classification, 66% of the areas are covered by very low to low vulnerability. These areas have some measures of protective cover which ranged from 7 m to 60 m (Oke *et al.*, 2018). Shallow groundwater is easily accessible by hand-dug wells, but the soil cover and thick vegetation in most cases may reduce infiltration and subsequent percolation. About 34% of the catchment of the Basin falls under moderate vulnerability zones. These areas are characterised by very shallow or no soil covers. Soil permeability is high under the PI moderate class areas.

The AVI shows larger areas (75%) classified as very high vulnerability (Figure 10.3). The very high vulnerability is due to the porous soil types of alluvium and sandstone, while 25% was assigned to the rest of the areas as high vulnerability. No area passed as very low or low to moderate vulnerability. The AVI vulnerability evaluation of the Dahomey Basin is strict in comparison to all the other methods. Pertaining to the DRAS-TIC evaluation of the Dahomey Basin vulnerability, low vulnerability areas covered 18% of the basin (Figure 10.2).

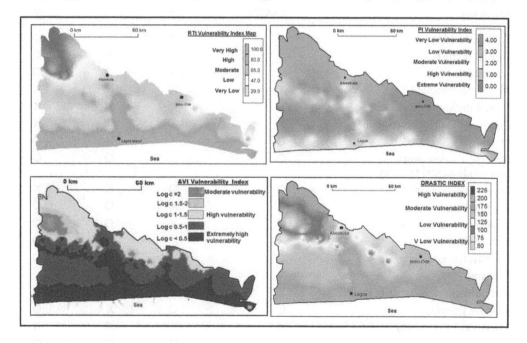

Figure 10.2 Comparison of vulnerability maps

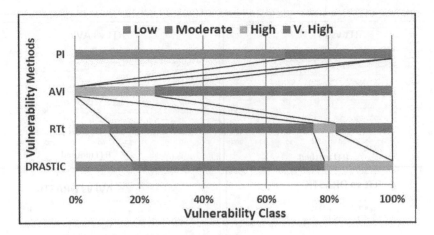

Figure 10.3 Comparison of percentage of vulnerability classes

Table 10.2 Vulnerability classification and range

Methods	Very low Vulnerability	Low Vulnerability	Moderate Vulnerability	High Vulnerability	Very High Vulnerability
RTt	12–29	29–47	47–65	65–83	83–100
DRASTIC	24–71	72–121	122–170	171–220	–
AVI	Log c >4	log c = 3–4	log c = 2–3	log c = 1–2	Log c <1
PI	>4–5	>3–4	>2–3	>1–2	>0–1

The areas with low vulnerability include high slope and depth-to-water, compacted soils and rock type. The majority of the areas in the basin were classified as moderate vulnerability with a value of 61%. However, in the other methods, the coast lines of the Atlantic Ocean and wetlands along the Ogun River covers about 21% high vulnerability class. The DRASTIC does not have a classification for very high or extreme vulnerability (Table 10.2).

3.3 Correlation plots

Further analysis between the methods was shown with correlation plots based on their normalised calculated vulnerability values. The highest correlation of 73% was observed between the AVI with DRASTIC method (Figure 10.4). This is followed with a 62% correlation between the RTt method with the DRASTIC method and 61% correlation between RTt with PI method (Table 10.3). The lowest correlation of 14% is between the DRASTIC methods with PI method. The low correlation between the PI and DRASTIC result from the duplication of parameters (Impact of vadose zone and aquifer media) in DRASTIC method which were simplified as lithology in the PI method and the low weight assigned to topography by DRASTIC.

The good correlation value between the AVI and DRASTIC method results from the few parameters considered in AVI method were all considered in the DRASTIC method.

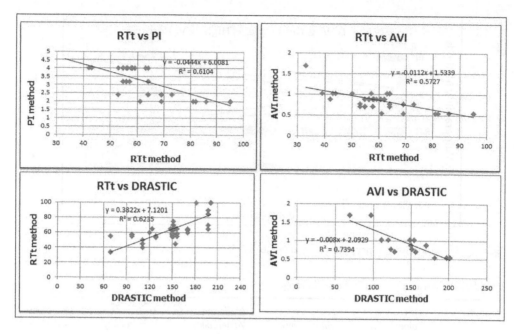

Figure 10.4 Cross plots of the four vulnerability methods

Table 10.3 Pearson correlation values of the four vulnerability methods

	AVI	PI	RTt	DRASTIC
AVI	1.0			
PI	0.52	1.0		
RTt	0.57	0.61	1.0	
DRASTIC	0.73	0.14	0.62	1.0

Furthermore, the AVI methods can directly be related to the physical properties of the vadose zone (Ross *et al.*, 2004) which DRASTIC over emphasised. The good correlation between RTt with DRASTIC resulted from both methods established based on the Delphi techniques system ratings (Aller *et al.*, 1987). The positive and negative correlations trends in Figure 10.4 result from the vulnerability methodology classification ranges in Table 10.2.

A normalised plot of the vulnerability method values of the borehole points is shown in Figure 10.5. The diagram shows the correlation of vulnerability degrees and the methods of DRASTIC, RTt, PI and AVI respectively. The differences in the diagram are the vulnerability class ranges used in the evaluation. RTt and DRASTIC were ranged up as vulnerability increases from low to high, while the AVI and PI methods were ranged down as the vulnerability increases from low to high (Table 10.2).

The AVI vulnerability method is the strictest of all the methods used in this study. This could be assumed as an overestimation of the groundwater vulnerability when compared to other methods. In a vulnerability study, it is better to overestimate as the AVI has done, than to underestimate the perceived risk. The aquifers in the PI moderate class is classified

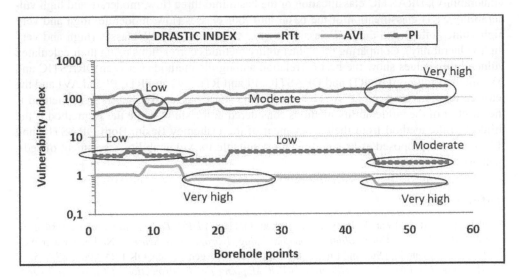

Figure 10.5 Normalised plot of vulnerability index

as high to extremely high by the RTt and AVI methods, and high by the DRASTIC method (Figure 10.5). Based on this, the PI method is the mildest of all the vulnerability methods applied in the evaluation of the Dahomey Basin. The areas in need of most protection are the upper aquifers of the basin especially the alluvium aquifers which are the nearest to the sea, highly populated and lowest depth to water level. All the vulnerability methods pointed out that these aquifers possess the greatest risk to contaminations and they ranged from borehole 45–58 (Figure 10.5). These aquifers need the most protection.

From the above analysis and comparison, it is obvious that the DRASTIC and RTt method looks most effective to evaluate shallow aquifers in sedimentary basins. They are also more reliable considering the following facts. (1) They considered more parameters (2) They have good correlations though AVI shows better correlation with DRASTIC than RTt but it considered less parameters (3) DRASTIC vulnerability classification range is flexible. Therefore, I propose the use of DRASTIC or modified DRASTIC methods for index method user and RTt method as best for those interested in evaluation based on travel time concept for sedimentary shallow aquifers vulnerability evaluations.

4 Conclusion

The study compared four vulnerability methodologies commonly used in groundwater assessments studies. The vulnerability methods were applied to the Dahomey Basin shallow aquifers of Nigeria. The comparisons include the methods parameter similarities and differences, vulnerability classifications of the basin and correlation plots of the methods. The PI method which is based on the European vulnerability concept contains the most assessments parameters while AVI consist of the list parameters. Other concept used in the formulating the methods is the point count system model, used in DRASTIC, PI and RTt methods.

Comparisons of the vulnerability classes as derived from the Dahomey Basin vulnerability maps, shows PI classification of the basin into two classes (low and moderate vulnerability), DRASTIC classification of the basin into three (low, moderate and high vulnerability), RTt classification of the basin into four classes (low, moderate, high and very high vulnerability) and the AVI classification of the basin into two classes (high and very high vulnerability). Comparing the vulnerability methods correlations using their calculated vulnerability values show the best correlation among the methods between DRASTIC and AVI method followed by RTt and DRASTIC, PI and RTt, AVI and RTt, PI and AVI and the least correlations between DRASTIC and PI. The study has shown that the AVI method is the strictest of the vulnerability methods considered in the study while the PI method is the mildest of the method from their assessment of the Dahomey Basin. Both DRASTIC and RTt method is proposed as best method most suitable for vulnerability assessment of sedimentary shallow aquifers.

References

Aller, L., Bennett, T., Lehr, J., Petty, R. & Hackett, G. (1987) *DRASTIC: A Standardised System for Evaluating Ground Water Pollution Potential Using Hydrogeologic Settings*. National Water Well Association, Dublin, Ohio and Environmental Protection Agency, Ada, OK.EPA-600/2-87-035.

COST Action 620 (2003) *Vulnerability and Risk Mapping for the Protection of Carbonate (Karst) Aquifer: Scope-Goals-Results*. European Commission Cost Action 620 Directorate-General Science, Research and Development L-2920: Luxembourg. pp. 1–42.

Daly, D., Dassargues, A., Drew, D., Dunne, S., Goldscheider, N., Neale, S., Popescu, C. & Zwhalen, F. (2002) Main concepts of the "European Approach" for (karst) groundwater vulnerability assessment and mapping. *Hydrogeology Journal*, 10(2), 340–345.

Faybishenko, B., Nicholson, T., Shestopalov, V., Bohuslavsky, A. & Bublias, V. (eds.) (2014) *Methodology of Groundwater Vulnerability and Protectability Assessment, in Groundwater Vulnerability*. Chernobyl Nuclear Disaster, John Wiley & Sons, Inc, Hoboken, NJ. doi:10.1002/9781118962220. ch4.

Gogu, R.C. & Dassargues, A. (2000) Current trends and future challenges in groundwater vulnerability assessment using overlay and index methods. *Environ. Geol*, 39(6), 549–559.

Goldscheider, N. (2002) *Hydrogeology and Vulnerability of Karst Systems, Examples from the Northern Alps and Swabian Alb*. PhD thesis, University of Karlsruhe. pp. 1–259.

Goldscheider, N., Klute, M., Sturm, S. & Hötzl, H. (2000) The PI method: A GIS-based approach to mapping groundwater vulnerability with special consideration of karst aquifers. *Z Angew Geol*, 46(3), 157–166.

Liggett, J.E. & Talwar, S. (2009) Groundwater vulnerability assessments and integrated water resource management. *Streamline Watershed Management Bulletin*, 13(1), 19.

Offodile, M.E. (2014) *Hydrogeology: Groundwater Study and Development in Nigeria*, 3rd edition. Mecon Geology & Engineering Services Ltd. pp. 485–514.

Oke, S.A. & Fourie, F. (2017) Guidelines to groundwater vulnerability mapping for sub-Saharan African. *Groundwater for Sustainable Development*, 5, 168–177. doi:10.1016/j.gsd.2017.06.007.

Oke, S.A., Vermeulen, D. & Gomo, M. (2016) Aquifer vulnerability assessment of the Dahomey Basin using the RTt method. *Environmental Earth Sciences*, 75(11), 1–9. doi:10.1007/s12665-016-5792.

Oke, S.A., Vermeulen, D. & Gomo, M. (2018) Intrinsic vulnerability assessment of shallow aquifers of the sedimentary basin of southwestern Nigeria. *Jàmbá: Journal of Disaster Risk Studies*, 10(1), a333. https://doi.org/10.4102/jamba.v10i1.333

Onwuka, M.O. (1990) *Groundwater Resources of Lagos State*. M.Sc thesis (Unpub), Dept. of Geology, University of Ibadan, Nigeria. pp. 144.

Ross, M., Richard, M., Lefebvre, R., Parent, M. & Savard, M. (2004) Assessing rock aquifer vulnerability using downward advective time from a 3D model of surficial geology: Case studies from the St. Lawrence Lowlands Canada. *Geofisica International*, 43(4), 591–602.

Van Stempvoort, D., Ewert, L. & Wassenaar, L. (1993) Aquifer vulnerability index: A GIS compatible method for groundwater vulnerability mapping. *Canadian Water Resources Journal,* 18, 25–37.

Vias, J.M., Andreo, B., Perles, M.J. & Carrasco, F. (2005) A comparative study of four schemes for groundwater vulnerability mapping in a diffuse flow carbonate aquifer under Mediterranean climatic conditions. *Environ Geol,* 47(4), 586–595.

Wachniew, P., Zurek, A.J., Stumpp, C., Gemitzi, A., Gargini, A., Filippini, M., Rozanski, K., Meeks, J., Kværner, J. & Witczak, S. (2016) Toward operational methods for the assessment of intrinsic groundwater vulnerability: A review. *Critical reviews in Environmental Science and Technology,* 46(9), 827–884. doi:10.1080/10643389.2016.1160816.

Vulnerability assessment of the Karst aquifer feeding Pertuso spring in Central Italy

G. Sappa, F. Ferranti & F.M. De Filippi

1 Introduction

Karst aquifers, including coastal aquifers, represent one of the most important freshwater resources for human life and economic activities, making up about 30% of the EU land mass (Foster *et al.*, 2013). Karst aquifers have complex hydrogeological characteristics, related to the high heterogeneity of hydraulic properties that make them different from other aquifers (Bakalowicz, 2005). These aquifers are characterised by a heterogeneous distribution of permeability due to conduits and voids developed by the dissolution of carbonate rocks (karst conduits, rutted fields, sinkholes, cave systems and ponors), frequently embedded in a less permeable fractured rock matrix (White, 1969). Flow velocities into a well-developed karst system are extremely fast and contaminants can quickly get to the saturated zone (Zwahlen, 2004). Thus, vulnerability assessment provides an important tool for delineating protection zones for water supply and highlighting areas where the aquifer is more susceptible to pollution introduced at the land surface (Marin *et al.*, 2015). With the development of Geographic Information System technology (GIS), there have been many new approaches for the evaluation of groundwater vulnerability to pollution (Doerfliger *et al.*, 1999; Hadžić *et al.*, 2015). Traditional approaches have strong limitations in vulnerability assessment of karst aquifer because the contaminant transport occurs mostly along preferential pathways (karst features) which makes modelling of karst systems challenging. A karst aquifer vulnerability technique has been developed which emphasises the karst formations outcropping in the study area. Two different methods for assessing intrinsic aquifer vulnerability were tested in a case study and their results compared. This paper presents the intrinsic vulnerability maps obtained by the application of SINTACS (Civita, 1994; Civita and De Maio, 2000) and COP method (Vias *et al.*, 2006) in the karst aquifer feeding the Pertuso Spring, in the Upper Valley of Aniene River (Central Italy), to identify the most vulnerable zone and the main processes that control the evolution of groundwater. The aim of this work is the comparison of both methods for intrinsic vulnerability mapping, to evaluate which one is more suitable in the study area.

2 Geological and hydrogeological setting

Pertuso Spring is an important karst spring located in the southern part of Latium Region, in the Upper Valley of Aniene River. This spring, with a discharge reaching up to 3 m^3/s, is the main water source for the southern part of Rome district (ACEA ATO2, 2005) and currently is feeding the Comunacqua hydroelectric power plant owned by ENEL group (Sappa and

Ferranti, 2013). The study area lies in the Upper Valley of the Aniene River (Figure11.1), in the southeast part of Latium Region (Central Italy) and covers an area of about 50 km² (Sappa and Ferranti, 2014). The Pertuso Spring hydrogeological basin is located between latitude 41°51' to 41°56'N and longitude 13°13' to 13°21'E and belongs to the Special Area of Conservation (SAC) of Aniene River Springs (EC Site Code IT6050029) established under Directive 92/43/EEC. The aquifer feeding Pertuso Spring (Figure 11.1) is mainly alternating granular limestone of Upper Cretaceous age and Triassic dolomite layers (Accordi and Carbone, 1988) locally covered by fluvial and alluvial deposits, Quaternary sediments, and Miocene clay and shale (Ventriglia, 1990) (Figure 11.1). The karst features in the aquifer feeding Pertuso Spring is relatively well developed at the surface of the carbonate outcrops, mainly in the Cretaceous limestone, where karren, sinkholes, groove and swallow holes can be observed (Accordi and Carbone, 1988; Damiani, 1990). This aquifer has a large storage capacity principally in the strongly fractured zones due to these karst features which provide low resistance pathways for groundwater that allow rapid flow of runoff to the saturated zone (White, 2002).

Recharge takes place from direct infiltration of rainfall, and the aquifer discharges through several springs (Table 11.1) located in the outcrop of Cretaceous limestone and Triassic dolomitic formations (Fig. 11.1).

This well-developed karst network drains the aquifer towards Pertuso Spring which is the main outlet of this karst aquifer (ACEA ATO2, 2005).

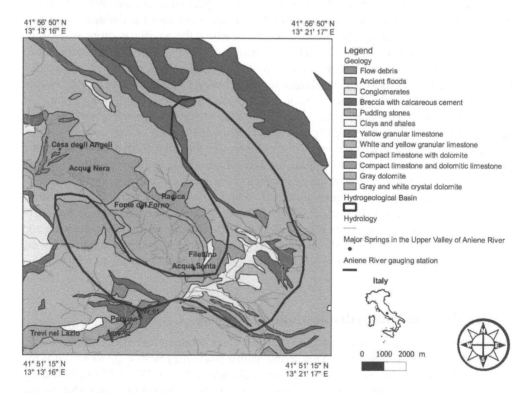

Figure 11.1 Geological map of the study area.

Table 11.1 Main springs in the Upper Valley of Aniene River (ACEA ATO2, 2005)

Spring	Altitude (m a.s.l.)	Average Discharge (l/s)
Acqua Santa	900	65
Acqua near	1030	80
Fonte del Forno	950	164
Cesa degli Angeli	940	200
Radica	1100	250
Pertuso	698	1400

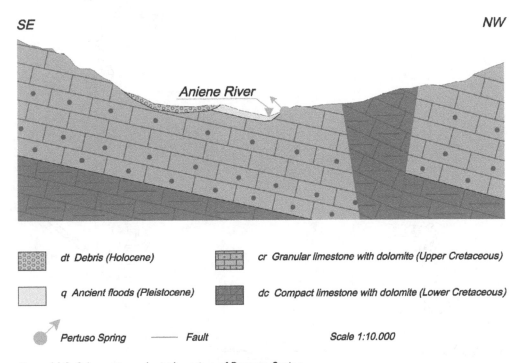

Figure 11.2 Schematic geological section of Pertuso Spring.

Pertuso Spring is located westward of Filettino (Figure 11.1), in the outcrop of limestone and dolomitic limestone of Upper Cretaceous age (Figure 11.2), upstream the town of Trevi nel Lazio, and flows into the Aniene River, close to the boundary of the carbonate hydrogeological system (Ventriglia, 1990).

Groundwater coming from Pertuso Spring, in the Upper Valley of Aniene River, comes from carbonate rocks. The most distinctive feature of Pertuso karst spring is the branching network of conduits that increase in size in the downstream direction (Figure 11.3). The largest active conduit drains the groundwater flow coming from the surrounding aquifer matrix, the adjoining fractures and the smaller nearby conduits. (Palmer, 1999; White and White, 1989). This conduit network is able to transmit large quantities of water through this karst aquifer (up to 3 m³/s) (Sappa *et al.*, 2016) (Figure 11.4).

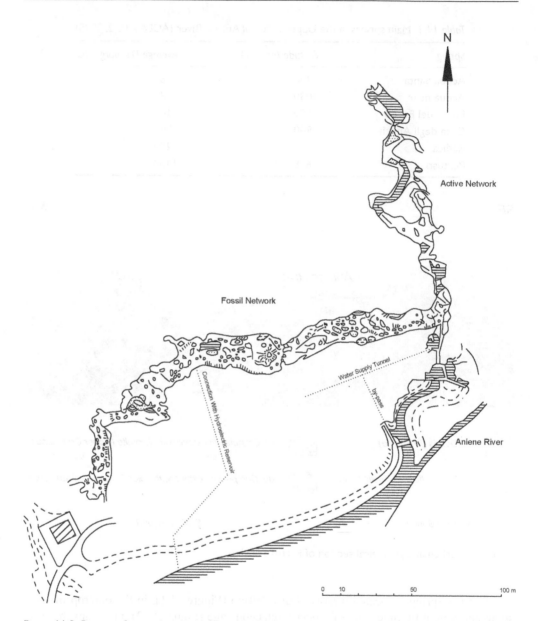

Figure 11.3 Pertuso Spring drainage gallery map with plan view of the development of karst drainage network (modified from ACEA ATO2, 2005).

Rainfall recharges the karst aquifer feeding Pertuso Spring mainly by fast infiltration in the saturated zone through karst features and fractures. Based on the record from 1990 to 1999 (ACEA ATO2, 2005), the highest Pertuso Spring discharge rate coincides with the largest rainfall event (Figure 11.4) and it is related to receipt of a stormwater sourced via rapid preferential flow through karst features and fractures.

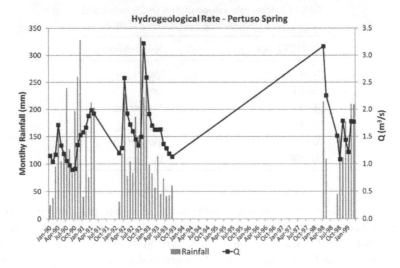

Figure 11.4 Temporal variation in Pertuso Spring discharge rate and rainfall in 1990–1999 period (Sappa and Ferranti, 2014).

3   Materials and methods

The aquifer vulnerability to pollution has been studied by applying the SINTACS method (Civita, 1994; Civita and De Maio, 2000) and COP method (Vias *et al.*, 2006).

3.1   Vulnerability assessment methods

3.1.1   SINTACS method

The SINTACS method has been developed to assess intrinsic groundwater vulnerability for hydrogeological and climatic impacts typical of the Mediterranean region, by the National Research Group for the Protection from Hydrogeological Disasters of the Italian National Research Council (Civita, 1994; Civita and De Maio, 2000).

SINTACS method uses seven parameters to estimate the vulnerability index, which ensures the best representation of the hydrogeological setting of the study area. The acronym SINTACS stands for the seven parameters included in this method: depth to groundwater (*S*), net recharge (*I*), effect of the vadose zone (*N*), effect of the soil media (*T*), aquifer media (*A*), hydraulic conductivity (*C*) and topographic slope (*S*). The model yields a numerical index that derived from scores and weights assigned to each parameter. A score (P_i), from 1 to 10, is assigned to each parameter, which are based on their relative effect on the aquifer vulnerability as defined by SINTACS method (Civita and De Maio, 2000). Weight multipliers (W_i), from 1 to 5, related to the specific anthropic and environmental conditions, are used for each parameter to enhance their importance in the vulnerability assessment (Corniello *et al.*, 2004).

The SINTACS vulnerability index (I_S) is defined by the scores of all the vulnerability parameters, multiplied by their respective weights (Equation 11.1).

$$I_S = \sum_{i=1}^{7} P_i \cdot W_i \tag{11.1}$$

Table 11.2 Intervals values of SINTACS index and corresponding vulnerability classes

SINTACS Index (I_s)	Vulnerability Class
0–24	Very Low
25–35	Low
36–49	Moderate
50–69	High
70–79	Very High
80–100	Extreme

The I_S index ranges from 26 to 260, but in order to facilitate interpretation of the results the data range has been normalised to 0–100 allowing six classes of vulnerability as shown in Table 11.2 (Civita, 1994; Civita and De Maio, 2000). SINTACS method assesses five vulnerability scenarios used to assign the correct weight multipliers to each parameter (Civita, 1994; Civita and De Maio, 2000): normal impact areas, relevant impact areas, areas with drainage from a superficial network, areas with karst features and fissured terrains.

3.1.2   COP Method

The COP method has been developed for the intrinsic vulnerability assessment of carbonate aquifers in the frame of the European COST Action 620 (COST Action 620, 2003). This method focuses on the role of karst characteristics of the aquifer, as factors that decrease the natural protection of the aquifer to pollution (Vias et al., 2006). According to this method, the natural degree of groundwater protection is related to three parameters: the properties of the overlying soils and the unsaturated zone (O factor), the protection due to diffuse or concentrated infiltration processes (C factor) and the variable climatic conditions (P factor). The COP method provides for the C factor for two different scenarios.

In Scenario 1, the C factor is calculated based on the parameters distance to the swallow hole (dh), distance to the sinking stream (ds) and the combined effects of slope and vegetation (sv) (Equation 11.2).

$$C = dh \cdot ds \cdot sv \tag{11.2}$$

Scenario 2 occurs in areas where the aquifer recharge is not through a swallow hole, so the C factor can be calculated on the bases of the parameters surface features (sf), slope (s) and the combined effects of slope and vegetation (sv) (Equation 11.3).

$$C = sf \cdot sv \tag{11.3}$$

The COP vulnerability index (I_{COP}) is obtain by multiplying the three factors (Equation 11.4).

$$I_{COP} = C \cdot O \cdot P \tag{11.4}$$

The COP vulnerability index classification defines five classes of vulnerability as shown in Table 11.3.

Table 11.3 Intervals values of COP index and corresponding vulnerability classes

COP Index (I_{COP})	Vulnerability Class
0–0.5	Very High
0.5–1	High
1–2	Moderate
2–4	Low
4–15	Very Low

3.2 Discretisation of the study area

The high discharge rate of Pertuso Spring (Figure 11.4) indicates the presence of karst conduits responsible of fast infiltration of rainfall in the saturated zone. Due to the relationship between rainfall and discharge rate, the preferential flow paths for groundwater in the study area have been inferred by the positions of the main karst features apparent from aerial photographs and satellite images. The traditional vulnerability assessment approaches require, before giving vulnerability rating, the partition of the study area into Finite Square Elements (EFQ) for which the Vulnerability Index is calculated. With the aim of emphasising the presence of karst features, SINTACS and COP methods have been separately applied to the Pertuso Spring hydrogeological basin (Figure 11.5): the first one consists of the direct recharge area, related to the presence of karst features responsible for fast infiltration of rainfall in the saturated zone, the second one to the whole basin.

The study area, about 50 km², has been divided into 22 polygons representative of outcropping lithology (Figure 11.5a) and 16 polygons related to karst features (Figure 11.5b). The aquifer vulnerability results, coming from both methods, are laid out, supported by QuantumGIS.

3.3 Required data

The intrinsic groundwater vulnerability assessment requires data related to the geological and hydrogeological settings of the aquifer, such as lithology, karst features, land use and land cover, soil, depth to groundwater, topography (slope) and climatic data (precipitation and temperature). The study is based on the background data coming from previous studies (Sappa and Ferranti, 2014) and the results of the first year of the Environmental Monitoring Plan, carried out in the karst aquifer of the Upper Valley of Aniene River, related to the catchment project of the Pertuso Spring. The data were collected from:

(i)   geological map on the scale 1:100.000,
(ii) Digital Elevation Model (DEM),
(iii) land use and land cover maps on the scale 1:10.000
(iv) climatic data (precipitation and temperature) for the 1992–2012 period in order to assess the average annual active recharge of the karst aquifer feeding Pertuso Spring,
(v)   aerial photographs and satellite images of the study area to identify the positions of the main karst features.

Data analyses were processed with QuantumGIS (version 2.12.2). Table 11.4 shows the major investigated layers used in the vulnerability assessment.

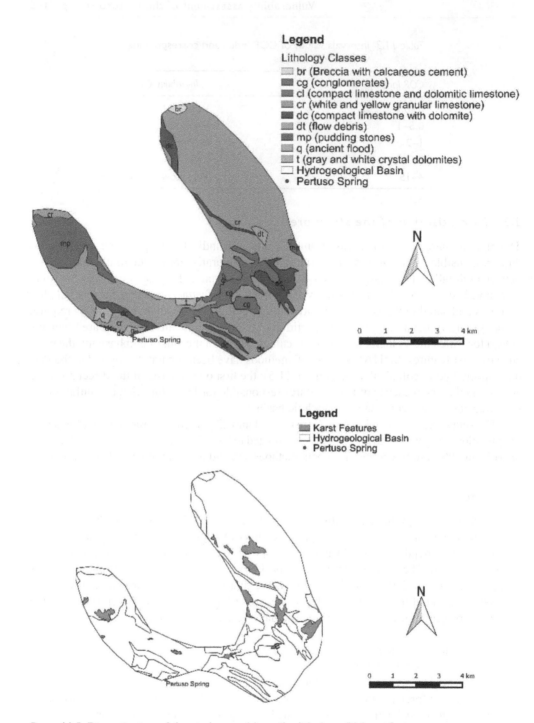

Figure 11.5 Discretisation of the study area: (a) aquifer lithology, (b) karst features.

Table 11.4 Data (type, reference and format) used for the vulnerability mapping (SINTACS and COP methods)

Data	Reference	Type	Vulnerability Method
Precipitation	Rome and Abruzzo Region Hydrographic Offices	Table	SINTACS-COP
Temperature	Rome and Abruzzo Region Hydrographic Offices	Table	SINTACS
Geology	Italian Geological Survey	Map	SINTACS-COP
Topography (slope)	Digital Elevation Model (DEM) provided by Sapienza University of Rome	Digital	SINTACS-COP
Land use and land cover	Lazio and Abruzzo Region-ISPRA (www.sinanet.isprambiente.it)	Digital	SINTACS-COP
Karst features	Aerial photographs and satellite images	Digital	SINTACS-COP

4   Results and discussion

4.1   Application of SINTACS method to the study area

4.1.1   Single parameter sensitivity analysis

The statistical summary of the seven rating parameter maps, used to evaluate the SINTACS index is provided in Table 11.5. As field data, referred to the water level contours for the study area, were not available, the depth to groundwater (Parameter *S*) was calculated by subtracting the altitude of Pertuso Spring (698 m a.s.l.), clearly inferior to the average altitude of the carbonate relief, from the height of the topographic surface. The topographic surface was obtained using the DEM (Digital Elevation Model) created from a 10 m spaced contour lines shapefile.

The topographic height values were extracted with QuantumGIS zonal statistics plugin from the DEM raster layer on the basis of the aquifer lithology and karst features vector polygons.

In the higher permeability area (karst features) the thickness of the unsaturated zone is greater than 350 m, while in the lower permeability area (aquifer lithology) the groundwater table surface has a maximum value of 1076.2 m and minimum value of 47.3 m. Thus, these values are classified into ranges according to the SINTACS method (Civita and De Maio, 2000) with rating from 0.3 to 1.8 as shown in Table 11.5.

The average annual active recharge of the Pertuso Spring hydrogeological basin (Parameter *I*) was calculated, according to the precipitation and evapotranspiration data by the application of the inverse hydrogeological water balance method (Civita and De Maio, 2000). Calculation of the effective infiltration is based on precipitation (*P*) and evapotranspiration (E_t) (Turc, 1954), combined with surface hydrogeological conditions, as contained in the potential infiltration index (c), according to the Equation 11.5:

$$I = (P - E_t) \cdot \chi \tag{11.5}$$

Table 11.5 A statistical summary of the single SINTACS parameters sensitivity analysis

Rating		Minimum	Maximum	Mean	Median	SD (Standard Deviation)
Aquifer	S	0.3	1.8	0.5	0.4	0.4
Lithology	I_A	4.5	9.0	6.9	6.9	1.7
	I_B	4.4	8.6	5.9	4.6	1.8
	I_C	5.7	9.0	8.3	8.9	1.1
	N	3.5	9.0	6.6	6.5	1.5
	T	10.0	10.0	10.0	10.0	0.0
	A	5.5	9.5	7.5	7.5	1.1
	C	1.6	10.0	3.4	4.1	1.9
	S	1.0	6.0	1.7	1.0	1.5
Karst	S	0.3	0.5	0.4	0.3	0.1
Features	I_A	4.5	8.8	5.9	4.6	2.0
	I_B	4.4	8.6	5.7	4.5	1.9
	I_C	5.2	9.0	7.0	6.2	1.6
	N	6.5	9.0	8.1	9.0	1.3
	T	10.0	10.0	10.0	10.0	0.0
	A	6.5	9.5	8.4	9.5	1.4
	C	1.6	4.1	3.3	4.1	1.2
	S	1.0	10.0	2.5	1.0	3.1

Table 11.6 Average annual active recharge for each rainfall scenario

		Scenario A	Scenario B	Scenario C
Meteorological Station		Vallepietra (RM)	Filettino (FR)	Carsoli (AQ)
Altitude (m a.s.l.)		825	1062	640
MPAM (mm/year)		1347	1624	1053
Active Recharge (Mm^3/year)	Aquifer Lithology	25.28	33.48	18.01
	Karst Features	1.17	1.54	0.82

Concerning the precipitation and evapotranspiration values, data registered in three meteorological stations located close to the study area were used in this study: Vallepietra (RM), Filettino (FR) and Carsoli (AQ). By analysing the average yearly rainfalls (MPAM) from 1992–2012 for each meteorological station, three different rainfall scenarios were applied to evaluate the average annual active recharge of the Pertuso Spring hydrogeological basin (Table 11.6): 1347 mm/year (Scenario A), 1624 mm/year (Scenario B) and 1053 mm/year (Scenario C).

According to the SINTACS method, in this study, with the aim of emphasising the high vulnerability of karst aquifer, the potential infiltration indexes χ were assigned on the basis of the ratings defined for areas, without any soil or with very scarce soil, choosing the values which make highest, the rates of the parameter I.

The results of the inverse hydrogeological water balance method applied to the karst aquifer feeding Pertuso Spring are reported in Table 11.6. These values are averaged over 20 years of observations (1992–2012). In the lower permeability area (aquifer lithology) the average annual active recharge ranged between 25.28 Mm³/year for Scenario A, 33.48 Mm³/year for Scenario B and 18.01 Mm³/year for Scenario C. In the karst features the average annual active recharge ranged between 1.17 Mm³/year for Scenario A, 1.54 Mm³/year for Scenario B and 0.82 Mm³/year for Scenario C. Thus, according to rating diagram of the parameter I (Civita and De Maio, 2000) a score ranging between 4.4 and 9.0 was assigned to the hydrogeological basin (Table 11.5).

The effect of the vadose zone (Parameter N) and the aquifer media (Parameter A) were obtained using official 1:100000 geological map, imported in digital format from the website of the Italian Geological Survey, geo-referenced and on-screen digitised to create a representation of the different geological polygonal layouts (Figure 11.5). The unsaturated zone of karst aquifer feeding Pertuso Spring consists mostly of granular limestone of Upper Cretaceous age and Triassic dolomites layers locally covered by fluvial and alluvial deposits, Quaternary sediments, pudding and Miocene clay and shale. Thus, the N rating was attributed between 3.5 and 9.0 to the lithotypes on the basis of lithotypes vs. N ratings diagram (Civita and De Maio, 2000) (Table 11.5). The elaboration of the data, related to the Parameter A followed the same course as the elaboration for parameter N. Ratings of aquifer media are based on permeability. The Cretaceous limestone is the most vulnerable to contamination and hence it was assigned the rating of 9.5 as the permeability is very high. The Triassic dolomite is the least vulnerable due to its impermeable nature and the rating is 5.5. Furthermore, where compact limestone, pudding stone, conglomerate and alluvial soil outcrop, which are permeable rock units, it was assigned a rating ranging between 6.5 and 7.0 (Table 11.5).

Soil use analysis of the study area was based on the maps of the CORINE Land Cover 2000 (ISPRA, 2006). In this karst area, because the soil is generally thin or absent, it was attributed the maximum rating (10.0) for the T parameter (Table 11.5). As field data, referred the hydraulic conductivity (Parameter C) were not available, it was assigned a score from 1 to 10 according to the horizontal bar chart of hydrogeological units of Civita and De Maio (2000) (Table 11.5). The parameter S refers to the slope percentage of the topographic surface. The slope was extracted with QuantumGIS from the Digital Elevation Model (DEM) in percentage, and reclassified into 11 categories ($\leq 2\%$, $3<S<4$, $5<S<6$, $7<S<9$, $10<S<12$, $13<S<15$, $16<S<18$, $19<S<21$, $22<S<25$, $25<S<30$ and $\geq 30\%$), to which were assigned weights (Table 11.5). The application of the SINTACS method for the vulnerability assessment of the karst aquifer feeding Pertuso Spring is summarised in Table 11.7, which shows the score of the different parameters used for the vulnerability mapping. Given the limited attenuation processes that make the karst aquifer more vulnerable to pollution, the SINTACS method introduces parameter weights that express the contribution of each parameter to the vulnerability. Thus, for the lower permeability area (aquifer lithology) each parameter has been multiplied by the fissured rocks weights set, while the higher permeability area (karst features) by the karst weights set.

Table 11.7 A statistical summary of the SINTACS vulnerability indexes

Rating	Aquifer Lithology			Kast Features		
	I_{SA}	I_{SB}	I_{SC}	I_{SA}	I_{SB}	I_{SC}
Minimum	35	32	37	42	40	42
Maximum	68	68	71	53	53	55
Mean	47	46	49	45	45	48
Median	46	44	50	45	45	48
SD (Standard Deviation)	6	7	7	3	3	4

4.1.2   Vulnerability assessment results

Following the procedure of SINTACS vulnerability method, the rated parameter maps were overlaid and three different final vulnerability maps were produced. These maps are represented in Figures 11.6, 11.8 and 11.10, respectively for annual rainfall of 1347 mm (Scenario A), 1624 mm (Scenario B) and 1053 mm (Scenario C). SINTACS indexes are divided into six classes and vary between the extreme values ranging from 0 to 100 (Civita and De Maio, 2000). Table 11.8 shows the results of SINTACS vulnerability method, expressed as

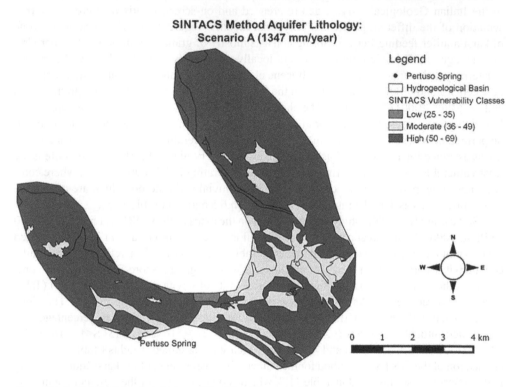

Figure 11.6 SINTACS Intrinsic Vulnerability Map (Scenario A).

Table 11.8 Comparison between the areas representing the SINTACS vulnerability classes obtained for each scenario applied

Vulnerability Class (%)	Scenario A		Scenario B		Scenario C	
	Aquifer Lithology	Karst Features	Aquifer Lithology	Karst Features	Aquifer Lithology	Karst Features
Very Low	–	–	–	–	–	–
Low	0.6	–	0.6	–	–	–
Moderate	17.6	96.0	26.5	96.0	13.7	48
High	81.7	4.0	72.9	4.0	85.9	52
Very High	–	–	–	–	0.4	–
Extreme	–	–	–	–	–	–

percentage distribution of vulnerability classes with respect to the total area of the hydrogeological basin and the area covered by karst formations.

The values of the SINTACS index vulnerability vary from 35 (Low) to 71 (Very High) (Table 11.7). Their spatial distribution distinguishes, within the Pertuso Spring hydrogeological basin, three main zones with different vulnerability degrees (Figures 11.6, 11.8 and 11.10), closely related to the hydrogeological setting of the study area.

The SINTACS aquifer vulnerability maps (Figures 11.6, 11.8 and 11.10) clearly show the dominance of high vulnerability classes in the northern part of the basin, while the southeastern and southwestern part are characterised by moderate vulnerability. There are no very low and extreme vulnerability classes. The low vulnerability class (Low: 25–35) covers less than 1% of the study area (Figures 11.7a and 11.9a) and is mainly due to the Triassic crystal dolomites lower permeability (Figures 11.6 and 11.8). This low degree of vulnerability is also influenced by the amount of precipitation; only the rainfall Scenarios A and B, related with the highest values of precipitations, respectively 1347 and 1624 mm/year, produce low vulnerability zones (Figures 11.6 and 11.8). Low vulnerability class is never assigned to karst features (Figures 11.7b, 11.9b and 11.11b).

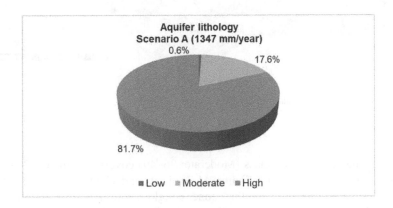

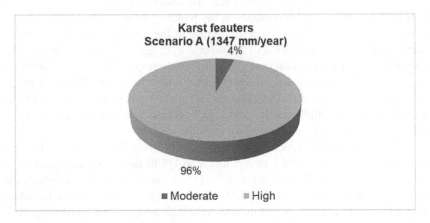

Figure 11.7 SINTACS vulnerability classes percentage distribution (Scenario A): (a) Aquifer lithology, (b) Karst features.

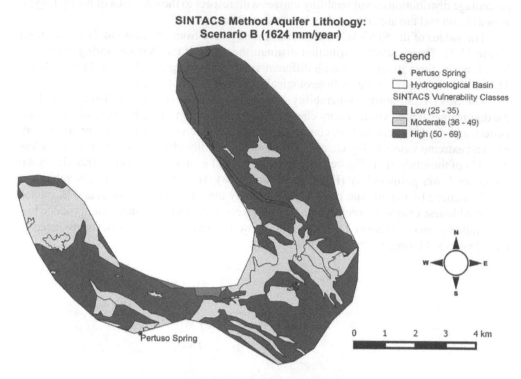

Figure 11.8 SINTACS Intrinsic Vulnerability Map (Scenario B).

The moderate vulnerability class (Moderate: 36–49) covers less than 25% of the Pertuso Spring hydrogeological basin (Figures 11.7a, 11.9a and 11.11a) and it is related to the presence of compact limestone, pudding stone, conglomerate and alluvial soil, which are permeable rock units. The highest rainfall scenario (Scenario B) has the major percentage of moderate classes (Figure 11.9a) mostly in the pudding stone outcropping in the western part of the basin (Figure 11.8). This result is due to the attribution of the rating of SINTACS parameter *I*. In the application of the inverse hydrogeological water balance method, the high rainfall rate of 1624 mm/year (Scenario B) is responsible for the highest values of infiltration (Table 11.6) and consequently for the lower rating of SINTACS parameter I (Figure 11.12). The high degree of vulnerability (High: 50–69) is due to the combination of karst granular limestone, high recharge and high hydraulic conductivity, allowing the rapid infiltration of pollutants, potentially coming from the surface.

The highest class of SINTACS vulnerability index (Very High: 70–79) (Figure 11.11) is present only in Scenario C and it covers less than 1% of the hydrogeological basin. This very high class is due to the lower precipitation (1053 mm/year) which decreases the score of parameter *I* (Figure 11.12). The SINTACS method applied to the karst features shows only two vulnerability classes: moderate and high, with a significant percentage difference between Scenario C and others (Figures 11.7b, 11.9b and 11.11b).

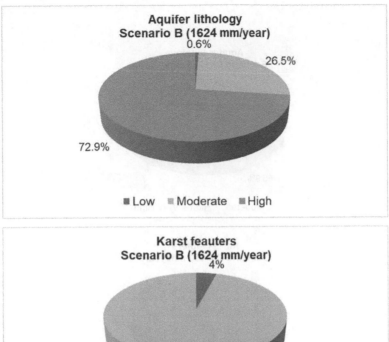

Figure 11.9 SINTACS vulnerability classes percentage distribution (Scenario B): (a) Aquifer lithology, (b) Karst features.

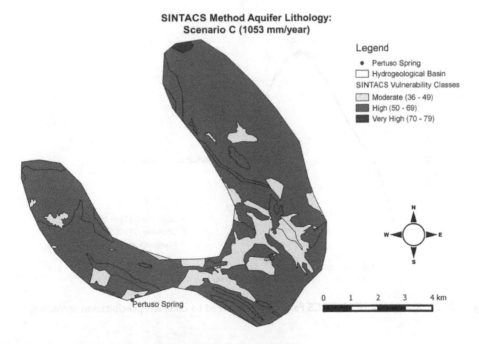

Figure 11.10 SINTACS Intrinsic Vulnerability Map (Scenario C).

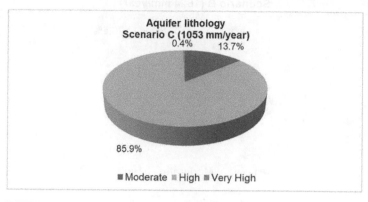

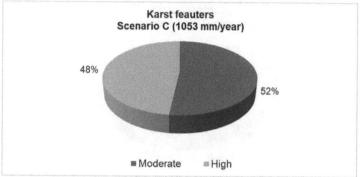

Figure 11.11 SINTACS vulnerability classes percentage distribution (Scenario C): (a) Aquifer lithology, (b) Karst features.

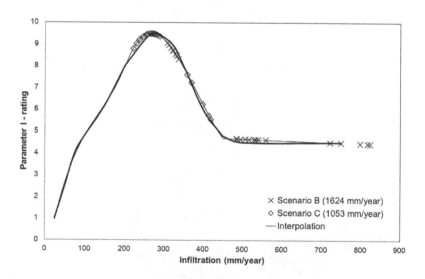

Figure 11.12 Rating plot of SINTACS Parameter I referred to different precipitation scenarios.

4.2   Application of COP method to the study area

4.2.1   Single parameter sensitivity analysis

The statistical summary of the seven rating parameter maps, used to evaluate the COP index is shown in Table 11.9.

For the evaluation of C factor, the COP method has two scenarios: the first zone includes the recharge area of karst features (Scenario 1) and the second one the rest of the area, where no surface karst features were identified (Scenario 2). The C score for Scenario 1 is obtained by multiplying the values of the parameter for slope and vegetation (sv) by those for the distances from recharge area to the swallow hole (dh) or the sinking stream (ds). The C score for Scenario 2 is evaluated by the multiplication of two main factors: vegetation and slope (sv) and surface features (sf). In this study, it was applied to Scenario 1 for the evaluation of C factor in the higher permeability area (karst features) and Scenario 2 in the lower permeability area (aquifer lithology). The slope and vegetation (sv) were extracted with QuantumGIS from the Digital Elevation Model (DEM) in percent and classified into 4 categories ($\leq$8%, 8% $\div$ 31%, 31%$\div$75% and >75%), which were assigned score respectively from 0.75 to 1. Based on the land cover maps provided, land use was divided into two main types, mainly no vegetation and vegetation. The distance to swallow holes (dh) consists of a series of buffer zones located at defined distances from fast recharge karst features (karren, sinkholes, groove and swallow holes). The recharge areas of these elements are made of the buffer around each identified karst features. It is assumed that the area located around each karst feature is characterised by a high vulnerability. Each buffer was attributed a respective factor dh. The distance to a sinking stream (ds) was assigned the value 1. The surface features parameter (sf) considers the geomorphological features specific to carbonate rocks and the presence or absence of any overlying layers (permeable or impermeable), which determine the importance of runoff and infiltration processes. The surface features parameter (sf) was assigned ratings from 0.5 to 1 according to surface geology. The C Factor ranges between 0 to 0.9 (Table 11.9) and is high ($\leq$ 0.2) in the higher permeability area (karst features), while is low (0.9) in non-karstic terrains, such as pudding stone, conglomerate, alluvial soil. The

Table 11.9 A statistical summary of the single COP parameters sensitivity analysis

Rating		Minimum	Maximum	Mean	Median	SD (standard deviation)
Aquifer	C	0.40	0.90	0.70	0.68	0.16
Lithology	O	1.00	5.00	2.90	3.20	1.40
	P_A	0.70	0.70	0.70	0.70	0.00
	P_B	0.80	0.80	0.80	0.80	0.00
	P_C	0.60	0.60	0.60	0.60	0.00
Karst	C	0.00	0.10	0.03	0.00	0.04
Features	O	2.00	5.00	3.25	3.00	1.34
	P_A	0.70	0.70	0.70	0.70	0.00
	P_B	0.80	0.80	0.80	0.80	0.00
	P_C	0.60	0.60	0.60	0.60	0.00

Table 11.10 A statistical summary of the COP vulnerability indexes

Rating	Aquifer Lithology			Kast Features		
	COP_A	COP_B	COP_C	COP_A	COP_B	COP_C
Minimum	0.32	0.36	0.27	0.00	0.00	0.00
Maximum	3.15	3.60	2.70	0.24	0.27	0.20
Mean	1.67	1.91	1.43	0.06	0.07	0.05
Median	1.42	1.62	1.22	0.00	0.00	0.00
SD (standard deviation)	1.00	1.14	0.86	0.09	0.10	0.08

O factor represents the overlying layers, namely the soil cover (O_S) overlying the bedrock lithology (O_L). The O_S factor represents the texture and thickness of the soil cover and was evaluated from the soil map. The O_L factor is representative of the unsaturated zone and is the product of the layer index and the degree of confinement (*cn*). The layer index is the product of the type of lithology and fracturing (*ly*) and the thickness of the unsaturated zone. The O_S and O_L factors were obtained using the geology map and drilling profiles of the Pertuso Spring karst aquifer (ACEA ATO2, 2005). The O factor ranges from 1 to 5, according with the thickness of the unsaturated zone (Table 11.9). The P Factor represents the climatic conditions in the catchment area and is the sum of two sub-factors (P_Q and P_I) defining, respectively, the amount and intensity of yearly precipitation. P_Q represents the amount of precipitation and ranges between 0.2 and 0.4. P_I represents the ratio of precipitation amount and number of rainy days.

The number of rainy days in the study area ranges from 91 to 107 days per year, based on the analysis of precipitation data for the three representative meteorological stations located near the study area in the years 1992–2012. This factor ranges between 0.6 and 0.8 (Table 11.9). The application of the COP method for the vulnerability assessment of the karst aquifer feeding Pertuso Spring is summarised in Table 11.10.

4.2.2   Vulnerability assessment results

The COP vulnerability index classification (Vias *et al.*, 2006) produced the vulnerability maps shown in Figures 11.13, 1.15 and 11.17, respectively for annual rainfall of 1347 mm (Scenario A), 1624 mm (Scenario B) and 1053 mm (Scenario C).

Table 11.11 shows the results of COP vulnerability method, expressed as percentage distribution of vulnerability classes with respect to the total area of the hydrogeological basin and the area covered by karst formations.

The COP vulnerability maps for each scenario show more differences in vulnerability degree, between Low and Very High, depending on the presence of specific conditions of factors O or C (Figures 11.13, 11.15 and 11.17).

The COP vulnerability maps (Figures 11.13, 11.15 and 11.17) show the dominance of high vulnerability classes (shades of red) in the northern part of the basin, while the southeastern and southwestern part are characterised by low vulnerability class. There are no very low vulnerability classes.

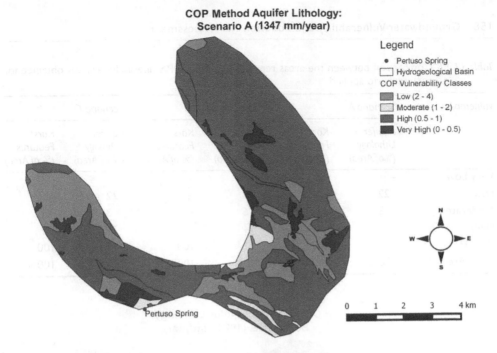

Figure 11.13 COP Intrinsic Vulnerability Map (Scenario A).

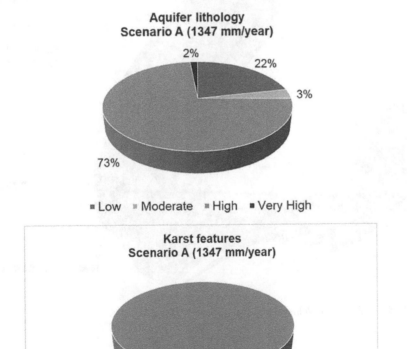

Figure 11.14 COP vulnerability classes percentage distribution (Scenario A): (a) Aquifer lithology, (b) Karst features.

Table 11.11 Comparison between the areas representing the COP vulnerability classes obtained for each scenario applied

Vulnerability Class	Scenario A		Scenario B		Scenario C	
	Aquifer Lithology (%of Area)	Karst Features (% of Area)	Aquifer Lithology (%of Area)	Karst Features (% of Area)	Aquifer Lithology (% of Area)	Karst Features (% of Area)
Very Low	–	–	–	–	–	–
Low	22	–	22	–	22	–
Moderate	3	–	6	–	3	–
High	73	–	71	–	71	–
Very High	2	100	1	100	4	100
Total Area (%)	100%	100%	100%	100%	100%	100%

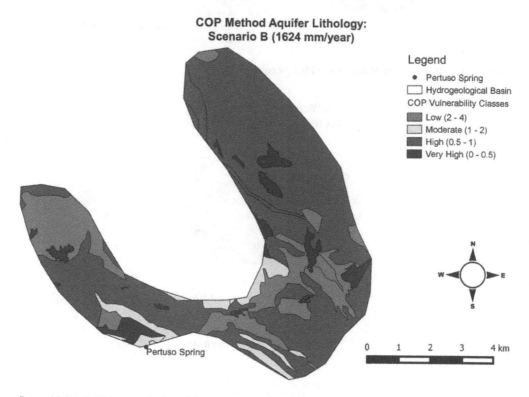

COP Method Aquifer Lithology:
Scenario B (1624 mm/year)

Legend

• Pertuso Spring
☐ Hydrogeological Basin
COP Vulnerability Classes
■ Low (2 - 4)
▨ Moderate (1 - 2)
■ High (0.5 - 1)
■ Very High (0 - 0.5)

Pertuso Spring

0   1   2   3   4 km

Figure 11.15 COP Intrinsic Vulnerability Map (Scenario B).

The karst features present in the study area are classified very high vulnerability (Very High: 0–0.5) (Figures 11.14b, 11.16b and 11.18b), due to the presence of swallow holes that decrease the residence time of the water in the unsaturated zone, reducing the potential attenuation capacity of the aquifer. This highest class of vulnerability index is due to the absence of an impermeable covering layer, which allowed rapid infiltration towards the

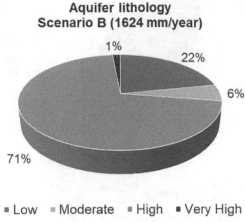

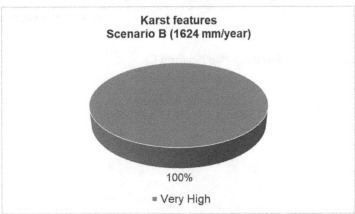

Figure 11.16 COP vulnerability classes percentage distribution (Scenario B): (a) Aquifer lithology, (b) Karst features.

saturated zone (*C* factor) (Figures 11.13, 11.15 and 11.17). The low vulnerability class (Low: 2–4) covers 22% of the study area (Figures 11.14a, 11.16a and 11.18a) in the outcropping of compact limestone, pudding stone, conglomerate and alluvial soil (Figures 11.13, 11.15 and 11.17). In these areas, the low permeability of cover layers and the gradient of the slope are responsible of an increase of *C* and *O* factors and consequently of a reduction of vulnerability. The moderate vulnerability class (Moderate: 1–2) covers only 6% of the Pertuso Spring hydrogeological basin (Figures 11.14a, 11.16a and 11.18a), where dolomitic limestone, compact limestone with dolomite and crystal dolomite outcrop decreasing the protective capacity assigned to the unsaturated zone by the *O* factor (Figures 11.13, 11.15 and 11.17). The COP method assigns the high vulnerability class (High: 0.5–1) to the major part of the study area (about 70 %) related to the karst Cretaceous limestone outcropping. The vulnerability maps obtained for each rainfall scenario are similar (Figures 11.13, 11.15 and 11.17). The differences depend on the amount of precipitation. Scenario C, relating to the lower value of precipitation (1053 mm/year), affects the *P* factor, which decreases to support the largest very high vulnerability zones (4%).

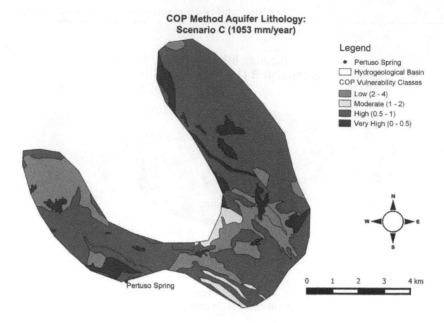

Figure 11.17 COP Intrinsic Vulnerability Map (Scenario C).

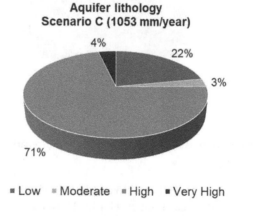

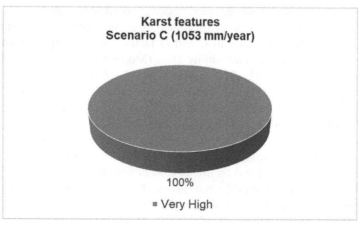

Figure 11.18 COP vulnerability classes percentage distribution (Scenario C): (a) Aquifer lithology, (b) Karst features.

4.3   Vulnerability methods comparison

Comparison between vulnerability results is based on the analysis of the three vulnerability maps obtained with SINTACS and COP method, applied for annual precipitation rates of 1347 mm (Scenario A), 1624 mm (Scenario B) and 1053 mm (Scenario C). SINTACS and COP methods output, respectively, three and four vulnerability classes, and provide relatively similar results for high vulnerability areas (Figures 11.19 and 11.20). They differ in moderate and very high-vulnerability classes. As far as it concerns vulnerability assessment of the karst aquifer feeding Pertuso Spring, SINTACS method defines mainly two classes, moderate and high vulnerability. SINTACS values fall within the range of 32–71, from low to very high vulnerability classes. There is no very low vulnerability class (Figure 11.19). COP values range from 0 to 3.6, including more vulnerability classes compared to the SIN-TACS method. In COP method four vulnerability classes were observed, but most of the study area is highly vulnerable. The very high vulnerability class is, however, rather important and mainly coincides with karst landscapes (Figure 11.20b). This is not reported by SIN-TACS method results (Figure 11.19). In SINTACS method moderate vulnerability zones are located where rocks, characterised by low permeability and high groundwater residence time (pudding stone, conglomerate, alluvial soil and crystal dolomite), outcrop, as they provide a natural contaminant attenuation within the unsaturated zone. The method assigns the high

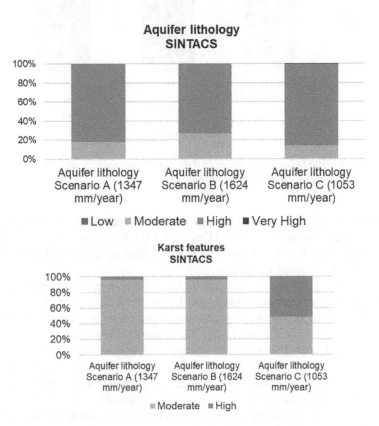

Figure 11.19 Comparison between the regrouped classes of vulnerability defined by the SINTACS method (a) Aquifer lithology (b) karst features.

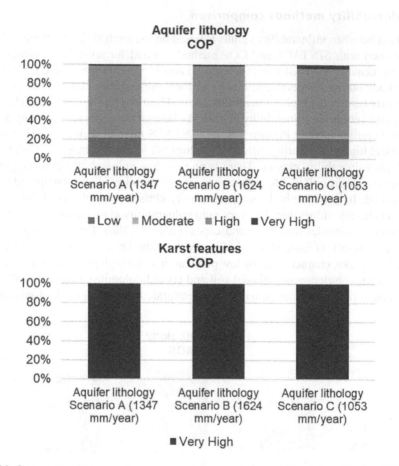

Figure 11.20 Comparison between the regrouped classes of vulnerability defined by the COP method (a) Aquifer lithology (b) karst features.

degree of vulnerability to the Cretaceous limestone outcrop, which allows the rapid infiltration of pollutants present on the surface (Figure 11.19a). Karst landscape shows a moderate vulnerability except for the lowest rainfall scenario, which assigns high degree to about 50% of the total karst area (Figure 11.19b).

COP vulnerability assessment assigns the lower level to the low permeability outcropping (pudding stone and conglomerate) and the moderate degree to the crystal dolomite (Figure 11.20a). The higher levels (high and very high) are assigned to the Cretaceous limestone and the karst formations (Fig 11.20b). Karst features do not always belong to the very high vulnerability class for both methods. For example, in SINTACS, high vulnerability was assigned to some karst features located in the eastern part of the basin, whereas COP method assesses them always as very highly vulnerable. This difference is due to the drainage effect from the overlying layers towards the swallow holes (Factor C). In conclusion, comparing vulnerability maps obtained using both methods, SINTACS seems to slightly underestimate the vulnerability and, unlike COP method, it presents a low sensitivity to the spatial variation of the hydrogeological parameters in a karst area.

4.4 *Validation by hydrogeochemical data*

Validation of the vulnerability assessments has been carried out on the results of the first year of the Environmental Monitoring Plan activities in the karst aquifer of the Upper Valley of Aniene River, related to the catchment project of the Pertuso Spring, which is going to be exploited to supply an important water network in the South part of Roma district.

Tracers, environmental and artificial, are powerful tools for validating hydrogeological conceptual models, so groundwater and surface water samples were collected from different gauging stations in four different monitoring campaigns, carried out in July and November 2014, and January and May 2015 and geochemical analysis used to better understand this catchment basin flow systems (Sappa *et al.*, 2017). As is to be expected in waters that circulate in karst aquifer, the Ca-Mg-HCO$_3$ facies is dominant in the Aniene River samples and result from dissolution of limestone and dolomite rocks which are abundant within the Triassic and Cretaceous formations which underlie much of the basin. The Ca-HCO$_3$ water type is the main hydrochemical facies, except for one sample (GW_01) which was similar to Ca-Mg-HCO$_3$ type, suggesting strong surface water-groundwater interaction.

The ionic ratio Mg/Ca (Figure 11.21) shows that water rock interaction processes and dissolution of carbonate minerals have influenced the groundwater chemistry in the study area (Gibbs, 1970). For surface water, the highest Mg/Ca ratios (0.5÷1) were found in the gauging station located upstream Pertuso Spring (SW_01).

The source of Mg^{2+} concentration values in Aniene River water upstream Pertuso Spring (SW_01) is the dissolution of Magnesium rich minerals in Triassic dolomites outcropping in the middle part study area (Sappa *et al.*, 2017). Thus, the increase in Mg/Ca ratio in surface water is due to the weathering of Triassic Mg-rich dolomite, where numerous karst springs take place (Figure 11.1).

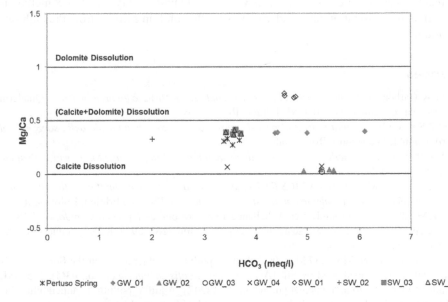

Figure 11.21 Plot of Mg/Ca versus HCO$_3$.

For groundwater, the high Mg/Ca ratios (~0.5) depend on the residence of water in the karst system, highlighting long residence time and enhanced weathering along the groundwater flow paths (White, 1988).

The karst aquifer feeding Pertuso Spring vulnerability assessment obtained by the application of SINTACS and COP methods are consistent with the results of the hydrogeochemical study. Areas with carbonate rocks are classified as zones with high and very high vulnerability levels, depending on their degree of karstification. The low and moderate vulnerability classes are located in the Triassic dolomite outcropping where less permeable materials are responsible for longer residence times within the karst aquifer.

5 Summary

The aim was to carry on a comparative evaluation of vulnerability assessment of the karst aquifer feeding Pertuso Spring, in the Upper Valley of Aniene River (Central Italy), by applying the SINTACS and the COP methods. The combined use of two vulnerability assessment approaches allows better understanding of mechanism of pollution vulnerability in the study area. The results of this study highlight that the vulnerability of the karst aquifer feeding Pertuso Spring mainly ranges from low to high for SINTACS method and from low to very high for COP method. The SINTACS and COP vulnerability maps show a high variability according to the environmental characteristics of the study area (e.g. geological, hydrogeological, morphological, climatic settings). The highest contribution to vulnerability was due to karst features such as karren, sinkholes, groove and swallow holes. This is due to the highly developed epikarst, high recharge and high hydraulic conductivity, which minimises the protective function of the unsaturated zone. The COP vulnerability assessment is more accurate than the SINTACS method. COP method shows more realistic results because it takes into account the role of the highly developed epikarst which minimises the protective function of the unsaturated zone. COP is the best method for the vulnerability assessment of karst features because it classifies them as very highly vulnerability. SINTACS method is more sensitive to precipitation rates and accounts for the dilution activity from the high rate of infiltration.

References

Accordi, G. & Carbone, F. (1988) *Carta delle litofacies del Lazio-Abruzzo ed aree limitrofe.* Quaderni de "La ricerca scientifica" 114(5), CNR Edition, Rome, Italy.

Acea ATO 2 S.p.A. (2005) *Studio idrogeologico-Proposta di aree di salvaguardia della sorgente del Pertuso.* Unpublished Report, Rome, Italy.

Civita, M. (1994) *Le carte della vulnerabilità degli acquiferi all'inquinamento. Teoria e Pratica.* Pitagora Edition, Bologna, Italy.

Civita, M. & De Maio, M. (2000) *SINTACS R5-Valutazione e cartografia automatica della vulnerabilità degli acquiferi all'inquinamento con il sistema parametrico.* Pitagora Edition, Bologna, Italy.

Bakalowicz, M. (2005) Karst groundwater: A challenge for new resources. *Hydrogeology Journal,* 13(1), 148–160. Available from: www.researchgate.net/publication/226575923_Karst_groundwater_A_challenge_for_new_resources.

Corniello, A., Ducci, D. & Monti, G.M. (2004) Aquifer pollution vulnerability in the Sorrento peninsula, southern Italy, evaluated by SINTACS method. *Geofísica Internacional,* 43(4), 575–581. Available from: www.researchgate.net/publication/26489077_Aquifer_pollution_vulnerability_in_the_Sorrento_peninsula_southern_Italy_evaluated_by_SINTACS_method.

COST Action 620 (2003) Vulnerability and risk mapping for the protection of carbonate (karst) aquifers. *European Commission, Directorate-General for Research.* Report EUR 20912, Luxemburg.

Damiani, A.V. (1990) Studi sulla piattaforma carbonatica laziale-abruzzese. Nota I. Considerazioni e problematiche sull'assetto tettonico e sulla paleogeologia dei Monti Simbruini, Roma. *Memorie descrittive Carta Geologica d'Italia,* 38, 177–206.

Doerfliger, N., Jeannin, P.Y. & Zwahlen, F. (1999) Water vulnerability assessment in karst environments: A new method of defining protection areas using a multi-attribute approach and GIS tools (EPIK method). *Environmental Geology,* 39(2), 165–176. Available from: https://link.springer.com/article/10.1007/s002540050446.

Foster, S., Hirata, R. & Andreo, B. (2013) The aquifer pollution vulnerability concept: Aid or impediment in promoting groundwater protection? *Hydrogeology Journal,* 21(7), 1389–1392. Available from: www.researchgate.net/publication/257471572_The_aquifer_pollution_vulnerability_concept_aid_or_impediment_in_promoting_groundwater_protection.

Gibbs, R.J. (1970) Mechanisms controlling world's water chemistry. *Science,* 170(3962), 1088–1090. Available from: https://web.viu.ca/krogh/chem301/Gibbs%20Science%201970%20world%20water%20chemistry.pdf.

Hadžić, E., Lazović, N. & Mulaomerović-Šeta, A. (2015) The importance of groundwater vulnerability maps in the protection of groundwater sources: Key study: Sarajevsko Polje. *Procedia Environmental Sciences. 7th Groundwater Symposium of the International Association for Hydro-Environment Engineering and Research (IAHR),* 25, 104–111.

ISPRA Ministero dell'Ambiente (2006) *Corine Land Cover.* [Online]. Available from: www.pcn.minambiente.it/viewer/.

Marin, A.I., Andreo, B. & Mudarra, A. (2015) Vulnerability mapping and protection zoning of karst springs: Validation by multitracer tests. *Science of the Total Environment,* 532, 435–446. Available from: www.researchgate.net/publication/278742853_Vulnerability_mapping_and_protection_zoning_of_karst_springs_Validation_by_multitracer_tests.

Palmer, A.N. (1999) Patterns of dissolution porosity in carbonate rocks. In Palmer, A.N., Palmer, M.V. & Sasowsky, I.D. (eds.) *Karst Modeling.* Karst Waters Institute Special Publication 5, Leesburg, VA. pp. 71–78.

Sappa, G. & Ferranti, F. (2013) Utilizzazione delle risorse idriche dell'Alta Valle dell'Aniene. *L'Acqua,* 2, 59–68. Available from: www.idrotecnicaitaliana.it/sommari/utilizzazione-delle-risorse-idriche-dellalta-valle-dellaniene/.

Sappa, G. & Ferranti, F. (2014) An integrated approach to the Enviromental Monitoring Plan of the Pertuso Spring (Upper Valley of Aniene River). *Italian Journal of Groundwater,* 3, 47–55. Available from: www.acquesotterranee.online/index.php/acque/article/view/60.

Sappa, G., Ferranti, F. & De Filippi, F.M. (2016) Hydrogeological water budget of the Karst aquifer feeding Pertuso spring (Central Italy). *International Multidisciplinary Scientific GeoConference Surveying Geology and Mining Ecology Management, SGEM,* 1, 847–854.

Sappa, G., Ferranti, F., De Filippi, F.M. & Cardillo, G. (2017) Mg^{2+}-based method for the Pertuso spring discharge evaluation. *Water,* 9(1), 67. Available from: www.mdpi.com/2073-4441/9/1/67.

Turc, L. (1954) Le bilan d'eaux des sols: relation entre les precipitations l'evaporation et l'ecoulement. *Annales Agronomiques,* 5, 491–595.

Ventriglia, U. (1990) *Idrogeologia della Provincia di Roma, IV, Regione orientale. Amministrazione Provinciale di Roma, Assessorato LL.PP.* Viabilità e trasporti, Rome, Italy.

Vias, J.M., Andreo, B., Perles, M.J., Carrasco, F., Vadillo, I., Jimenez, P. & Zwalhen, F. (2006) Proposed method for groundwater vulnerability mapping in carbonate (karstic) aquifers: The COP method, application in two pilot sites in Southern Spain. *Hydrogeology Journal,* 14(6), 912–925. Available from: www.researchgate.net/publication/225673265_Proposed_method_for_groundwater_vulnerability_mapping_in_carbonate_karstic_aquifers_The_COP_method.

White, W.B. (1969) Conceptual models for carbonate aquifers. *Groundwater,* 7, 15–21. Available from: https://onlinelibrary.wiley.com/doi/abs/10.1111/j.1745-6584.1969.tb01279.x.

White, W.B. (1988) *Geomorphology and Hydrology of Karst Terrains*. Oxford University Press, New York.

White, W.B. (2002) Karst hydrology: Recent developments and open questions. *Engineering Geology*, 65(2–3), 85–105. Available from: www.researchgate.net/publication/222667368_Karst_hydrology_ Recent_developments_and_open_questions.

White, W.B. & White, E.L. (1989) *Karst Hydrology-Concepts from the Mammoth Cave Area*. Van Nostrand Reinhold, New York. p. 346.

Zwahlen, F. (ed.) (2004) *Vulnerability and risk mapping for the protection of carbonate (karst) aquifers: Final Report COST Action 620*. European Commission, Directorate-General for Research, Brüssel, Luxemburg. Available from: https://publications.europa.eu/en/publication-detail/-/publication/ be3c99bf-1a0a-4213-b35d-c3faffcd355b.

Groundwater vulnerability mapping – examples of different national approaches

Chapter 12

The groundwater contamination potential risk evaluation

An all-country wide approach for protection planning

M.V. Civita

1 Introduction

The hydrogeological risk *sensu stricto*, i.e. is the most difficult of anthropogenic risk to be evaluated. As the risk (R) may be described by the expression:

$$R = HT \times V_u \times V_a \tag{12.1}$$

in which H_T is the hazard, V_u is the vulnerability of the aquifer subject to risk, and V_a is the social & economic value of the resource (Civita, 2000), is immediately clear that reliable values of H_T are practically impossible to gain when the contamination processes happen within the subsoil section of the environmental system. H_T, indeed, is a statistical factor including the probability that an impact event may happen in a given time. Normally, in the overdeveloped country too, only sometimes historical data series underground pollution are available to quantify the statistical value of the impact (frequency, magnitude, return time). Only when an impact event is monitored for a long time, as happen for instance in some polluted sites included in the CERCLA priority list, time/space series of contamination in progress are available, moreover before the impact is happened. Since sixties, several *site analysis* and *potential contamination risk techniques* was proposed to assess punctual contamination source (LeGrand, 1964, 1983; Born and Stephensons, 1969; Hughes *et al.*, 1971; Pavoni *et al.*, 1972; Palmquist and Sendlein, 1975; USEPA [OFRNARA], 1994; Civita and Zavatti, 2006). Anyway, the problem of protecting large aquifers against the concomitance action of punctual and diffuse contamination sources seem insoluble but very urgent in presence of the rise of consumption and the fall of quantity and quality of the groundwater resources.

Early in the nineties, the scientific community together with a number of decision makers understood the urgency to protect natural and environmental resources. They agree that an adequate level of knowledge scientifically organised allows an accurate planning and development of *an environment system* by ruling and directing the effective development process without stopping it.

In the main sector of groundwater resource (GWR) for human consumption, since the early seventies, a really new approach was born strongly oriented toward a *territory wide* assessment of the factors affecting the GWR contamination potential whit the task to protect *the resource* and not only the *tapping work* points as *risk targets*. The target was pointed out to the prevision and prevention of the events and not only to the groundwater reclamation action planning.

The above said approach allows the assessment of intrinsic vulnerability of aquifers subjecting to a less or more developed territory, which expression is a thematic document, the *groundwater (aquifer) vulnerability to contamination map*. In this thematic map, the intrinsic vulnerability together with the groundwater flux trend and the geographic position and social-economic importance of the withdrawal points are compared to the position and hazard class of the potential contamination sources (PCS) existing in the area. The map then become a forecasting tool and, via the planning process, a prevention tool and an identifier of structural and non-structural action priority list.

2   History

The evaluation of the specific vulnerability of an aquifer should be made case by case, taking into account all the chemical and physical features of each single contaminant that is present (or of a group of similar contaminants), the type of source (punctual or diffused), quantity, means and rates of contaminant applications (Andersen and Gosk, 1987; Foster and Hirata, 1988; Bachmat and Collin, 1987). This approach, although scientifically valuable and adequate for the case of the evaluation of a potential contamination of a CSC in small areas, is quite impracticable where the goal is the assessment of aquifer vulnerability for large areas or when it is carried out as part of contamination prevention and aquifer protection planning. In the last 30 years, a number of techniques have been developed for the general treatment of data (Table 12.1).

These techniques vary considerably, according to the physiography of the tested areas, to the quantity and quality of the data, and to the aim of the study.

In Italy, an important national program of applied research on the topic was developed since 1984, funded by Civil Defence Ministry and entrusted to National Council of Research. The work of almost 100 researchers distributed all country wide in 20 task-units has covered almost 135,000 km^2 (in front of 150,000 km^2 under contamination risk), developing and testing new approaches and methodologies in very different hydrogeological scenarios, from the large alluvial plain to the fissured rocks built mountains.

The intrinsic (i.e. natural) vulnerability of an aquifers to contamination is the specific susceptibility of aquifer systems, in their various parts and in the various geometric and hydrodynamic settings, to ingest and diffuse fluid and/or hydro-vectored contaminants, the impact of which, on the groundwater quality, is a function of space and time (Civita, 1987). It depends of three main factors: the ingestion and time of travel (TOT) of contaminant fluid; the flow dynamics in saturated zone; and the residual concentration of contaminant reached the groundwater.

At the first time (1985), the Group began to work using a derived French method (Albinet and Margat, 1970) coming from an adjusted face to the Italian geological features and the whole environment (Francani and Civita, 1988).

The same method for intrinsic vulnerability (Table 12.2) was presented (Civita, 1990b) coupled with a complete legend of pollution CSC (Contamination Spreading Centre) and situations of spreading contaminants. This pollution legend allows a first integrated (superimposed) vulnerability map. During the activity period of the Group 66 maps where prepared ranging from 10,000–1 to 250,000–1 scales, with a large frequency of 25,000–1, using this method (Civita and De Maio, 2000).

A *rating system* was concerned in this period, the very simple GOD (Foster and Hirata, 1988) poorly used in Italy.

Table 12.1 Methods for the assessment of intrinsic aquifer vulnerability contamination and relative basic information (from Civita, 1994). Explanation: AR = Analogical relations; HCS = Hydrogeological complex and setting; MS = Matrix system; RS = Rating system; PCSM = Point Count System Model (Referees in Civita 1994).

METHODOLOGY

NATURAL INFORMATION

REFERENCE AND/OR NAME	TYPE	PRECIPITATION RATE & CHEMICAL COMPOSITION	TOPOGRAPHIC SURFACE & SLOPE VARIABILITY	SURFICIAL STREAMFLOW & NETWORK DENSITY	THICKNESS, TEXTURE & MINERALOGY	EFFECTIVE MOISTURE	PERMEABILITY	PHYSICAL & CHEMICAL PROPERTIES	AQUIFER CONNECTIONS TO SURFICIAL WATERS	NET RECHARGE	HYDROGEOLOGIC FEATURES OF INS. ZONE	DEPTH TO WATER	PIEZOMETRIC LEVEL CHANGES	AQUIFER HYDROGEOLOGIC FEATURES	AQUIFER HYDRAULIC CONDUCTIVITY
					CHARACTERISTICS OF SOIL										
Albinet and Margat (1970) and BRGM (1970)	HCS								•		•	•	•	•	•
Vrana (1968) and Olmer and Rezac (1974)	HCS										•	•	•	•	•
Fenge (1976)	RS				•					•	•			•	•
Josopait and Swerdtfeger (1976)	HCS									•	•			•	•
Vierhuff et al. (1981)	HCS										•			•	
Zampetti (1983) and Fried (1987)	AR										•	•	•	•	•
Villumsen et al. (1983)	RS				•						•				
Haertle' (1983)	MS				•										
Vrana (1984)	HCS	•										•			
Subirana Asturias and Casas Ponsati (1984)	HCS								•		•	•		•	•
Engelen (1985)	HCS								•		•	•		•	

(Continued)

Table 12.1 (Continued)

REFERENCE AND/OR NAME	METHODOLOGY	NATURAL INFORMATION			CHARACTERISTICS OF SOIL										
	TYPE	PRECIPITATION RATE & CHEMICAL COMPOSITION	TOPOGRAPHIC SURFACE & SLOPE VARIABILITY	SURFICIAL STREAMFLOW & NETWORK DENSITY	THICKNESS, TEXTURE & MINERALOGY	EFFECTIVE MOISTURE	PERMEABILITY	PHYSICAL & CHEMICAL PROPERTIES	AQUIFER CONNECTIONS TO SURFICIAL WATERS	NET RECHARGE	HYDROGEOLOGIC FEATURES OF INS. ZONE	DEPTH TO WATER	PIEZOMETRIC LEVEL CHANGES	AQUIFER HYDROGEOLOGIC FEATURES	AQUIFER HYDRAULIC CONDUCTIVITY
Zaporozec (1985)	RS				•	•	•	•	•		•	•		•	•
Breeuwsma et al. (1986)	HCS				•	•	•	•		•	•	•		•	•
Sotornikova and Vrba (1987)	RS						•								•
Ostry et al. (1987)	HCS				•							•	•	•	
Goossens and Van Damme (1987)	MS				•			•				•		•	
Carter et al. (1987) Palmer (1988)	MS				•		•	•				•		•	
Marcolongo and Pretto (1987)	RS														
Marcolongo and Pretto (1987)	AR					•				•		•			
GOD Foster (1987), Foster and Hirata (1988)	RS									•	•	•		•	
Schmidt (1987)	RS										•	•			
Trojan and Perry (1988)	PCSM	•	•		•		•		•	•	•	•			
GNDCI Natural (Civita, 1990a)	HCS										•	•		•	
DRASTIC Aller et al. (1987)	PCSM		•		•				•		•	•		•	•
SINTACS (Civita and De Maio (1997, 2000)	PCSM		•	•	•				•	•	•	•		•	•

Table 12.2 Standards of Italian hydrogeologic/vulnerable settings (GNDCI-CNR basic method)(after Civita, 1990a)

Vulnerability degrees	Hydrogeologic complexes and setting features
Extremely high	Unconfined (water table) aquifer in alluvial deposits: streams that freely recharge the groundwater body; well or multiple well systems that drawdown the water table to under the stream level (forced recharge). Aquifer in carbonate (and sulfate) rocks affected by completely developed karst phenomena (holokarst with high karst index [ki]).
Very high	Unconfined (water table) aquifer in coarse to medium-grained alluvial deposits, without any surficial protecting layer. Aquifer in highly fractured (high fracturing index [fi]) limestone with low or null ki and depth to water <50m.
High	Confined, semi-confined (leaky) and unconfined aquifer with impervious (aquaculture) or semi-pervious (*aquitard*) superficial protecting layer. Aquifer in highly fractured (high fracturing index) limestone with low or null ki and depth to water >50m. Aquifer in highly fractured (but not cataclastic) dolomite with low or null ki and depth to water <50m. Aquifer in highly clivated volcanic rocks and non-weathered plutonic igneous rocks with high fi.
Medium	Aquifer in highly fractured (but not cataclastic) dolomite with low or null ki and depth to water >50m. Aquifer in medium to fine-grained sand. Aquifer in glacial till and prevalently coarse-grained moraines.
Medium – Low	Strip aquifers in bedded sedimentary sequences (shale-limestone-sandstone *flysch*) with layer by layer highly variable diffusion rates. Multi-layered aquifer in pyroclastic non indurated rocks (tuffs, ash, etc.): different diffusion degrees layer by layer close to the change in grain size.
Low	Aquifer in fissured sandstone or/and non-carbonate cemented conglomerate. Aquifer in fissured plutonic igneous rocks. Aquifer in glacial till and prevalently fine-grained moraines. Fracture network aquifer in medium to high metamorphism rock complexes.
Very low or null	Practically impermeable (*aquifuge*) marl and clay sedimentary complexes (also marly *flysch*): contamination directly reaches the surface waters. Practically impermeable (aquifuge) Fine-grained sedimentary complexes (clay, silt, peat, etc.) contamination directly reaches the surface waters. Meta-sediment complexes or poorly fissured, highly tectonized clayey complexes low metamorphism complexes, almost aquifuge: contamination directly reaches the surface waters.

Before 1988, the DRASTIC method (Aller *et al.*, 1987) was envisaged and concerned as an important new PCSM (Point Count System Model). It collected 7 basic parameters and 2 different pound strings. But it was criticized in some parts (Civita, 1994) and was considered poorly adapted to Italian very variable landscape and GIS application.

Furthermore, some DRASTIC parameters allows to use base-data impossible to have in Italy. The maps produced using DRASTIC were only 5, although some bad variations of the method were appointed (i.e. DAC, VIR, DRASTIC.MC, AVI and so on). The reader can refer to Civita (1994) and to Vrba and Zaporozec (1995) for an exhaustive discussion of the previously mentioned methods.

The research experience as a whole and the GIS-based application (Geographical Information System) was performed in the next years following a new PCSM usable in Italy's complex environment as *a planning scheme*. The new methodology was called SINTACS coming from the Italian names of the factors that are involved, i.e. *S oggiacenza* (depth to groundwater); *I nfiltrazione* (effective infiltration); *N on saturo* (unsaturated zone attenuation capacity); *T ipologia copertura* (soil/overburden attenuation capacity); *A cquifero* (saturated zone characteristics); *C onducibilità* (hydraulic conductivity*); *S uperficie topografica* (topographic surface slope). These factors, of course, are the same used in DRASTIC model, therefore, some people affirmed that SINTACS is a derivation of DRASTIC. But, for 5 factors a lot of fundamental changes was introduced and the others were performed to Italian landscapes, 6 waiting strings was introduced to represent all the contamination situation. A GIS – based application was performed to allow directly use and updating and subsequent correcting data in every municipal bureau and environment Agency. A number of maps (51) from 10,000^{-1} to 400,000^{-1} have been compiled since 2001 in every type of Italy's all country environment.

As has been widely verified from a comparison of several different approaches applied to the same sample-area (Civita, 1994), the choice of the method that is most suitable to build a vulnerability map for a certain area should initially depend on a strictly realistic evaluation of the number, distribution and reliability of the available (and/or surveyable) data.

SINTACS (Table 12.3) is now the complex methodology valid all country wide that take place in the most important state norm for protection of water against pollution (Low Decree # 152/99 and coordinate norms). The methodologies and the GIS – based application were published in a specific Guide – Line by the Italian Environment Agency (ANPA, 2001).

Table 12.3 Short description of the parameters and related rating graphs for PCSM SINTACS. For a complete description, see Civita and De Maio (2000)

DESCRIPTION	RATING DEFINITION
S Depth to groundwater: is defined as the depth of the piezometric level (both for confined or unconfined aquifers) with reference to the ground surface and it was a great impact on the vulnerability because its absolute value, together with the unsaturated zone characteristics, determine the time of travel (tot) of a hydro-vectored or fluid contaminant and the duration of the attenuation process of the unsaturated thickness, in particular the oxidation process due to atmospheric O_2. The SINTACS rating of depth-to-groundwater therefore decreases with an increase of the depth, i.e. with an increase of the thickness of the unsaturated zone within the range 10 ÷ 1.	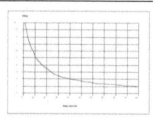
I Effective infiltration action: The role that the effective infiltration plays in aquifer vulnerability assessment is very significant because of the dragging down surface of the pollutant but also their dilution, first during the travel through the unsaturated zone and then within the saturated zone. Direct infiltration is the only or widely prevalent component of the net recharge in all the areas where there are no interflow linking aquifers or surficial water bodies or no irrigation practices using large water volumes.	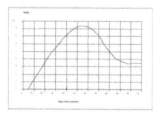

N Unsaturated zone attenuation capacity: The unsaturated zone is the "second defense line" of the hydrogeologic system against fluids or hydro-vectored contaminants. A four dimension process takes place inside the unsaturated thickness in which physical and chemical factors synergically work to promote the contaminant attenuation. The unsaturated zone attenuation capacity is assessed starting from the hydrolithologic features (texture, mineral composition, grain size, fracturing, karst development, etc.).

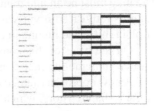

T Soil/overburden attenuation capacity This is the "first defense line" of the hydrogeologic system: several important processes take place inside the soil that built up the attenuation capacity of a contaminant travelling inside a hydrogeologic system and therefore in aquifer vulnerability assessment and mapping. Soil is identified as an open, three-phase, accumulator and transformer of matter and an energy sub-system which develops through the physical, chemical and biological alterations of the bottom lithotypes and of the organic matter that it is made up of.

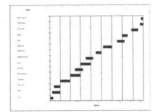

A Hydrogeologic characteristics of the aquifer: In vulnerability assessment models, the aquifer characteristics describe the process that takes place below the piezometric level when a contaminant is mixed with groundwater with a loss of a small or more relevant part of its original concentration during the travelling through the soil and the unsaturated thickness. Basically these processes are: molecular and cinematic dispersion, dilution, sorption and chemical reactions between the rock and the contaminants.

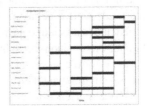

C Hydraulic conductivity range of the aquifer: Hydraulic conductivity represents the capacity of the groundwater to move inside the saturated media, thus the mobility potential of a hydro-vectored contaminant which as a density and viscosity almost the same as the groundwater. In the SINTACS assessment context, the hydraulic gradient and the flux cross-section being equal, this parameter determines, the aquifer unit yield and flow velocity that go toward the effluences or the tapping work that indicates the of risk targets.

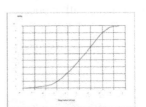

S Hydrologic role of the topographic slope: The topographic slope is an important factor in vulnerability assessment because it determines the amount of surface runoff that is produced, the precipitation rate and displacement velocity of the water (or a fluid and/or hydro-vectorable contaminant) over the surface being equal. A high rating is assigned to slight slopes i.e. to surface zones where a pollutant may be less displaced under gravity action or even stop in the outlet place favouring percolation. The slope may be a genetic factor due to the type of soil and its thickness, and can indirectly determine the attenuation potential of the hydrogeologic system.

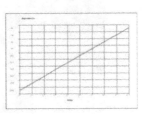

3 The Italian "combined" approach

The map must be an integrant part of a *land planning scheme* for any order and degree of the administrative territory: it cannot depend on the morphology as it must cover a wide mixture of plain, hilly and mountain areas, as can be found throughout Italy.

On the basis of this consideration, it was realized that it is impossible to elaborate an aquifer vulnerability map using one single method.

A new approach (named *combined approach*) was studied and tested, for use in any part of the Italian territory, which was based on the overlapping of two different methodologies:

- areas (1 in Fig.12.1), where the amount and reliability of data, measurements, tests and analysis can be considered to be sufficient for the mapping scale
- homogeneous areas zoning, based on the survey of hydrogeologic complexes, characteristics and settings (HCS), to be used in mountainous and hilly areas (2 and 3 in Fig. 12.1) where a scarcity or underground information is normal (GNDCI-CNR *Basic [Natural] Method*).

From what has been seen, in many areas where it is necessary to cover vast areas identified by administrative (i.e. Municipalities, Provinces, Regions) or physical boundaries (interregional watershed) with a Vulnerability Map, the parametric models that have been set up cannot be applied due to a lack of data at those points where the terrain changes from a plain morphology to a hilly or mountainous area. In these situations, in the past, a simple method was chosen that was able to perform a less refined and detailed evaluation, but which however was applied to many land and environmental problems connected to the contamination of aquifers with good results.

The experience gained over recent years has led to a reconsideration of the methodological problem: why renounce the detail that can be offered by a Point Count System Models in areas with moderate relief where the majority of the CSCs And the DCAs (Diffused

Figure 12.1 Geographical position of the test sites used to test the subdivision of SINTACS index in several vulnerability degrees. A parametric method (a highly advanced PCSM – i.e. SINTACS Release 5 [Civita and De Maio, 2000]), which has been improved for plain, piedmont and mountains.

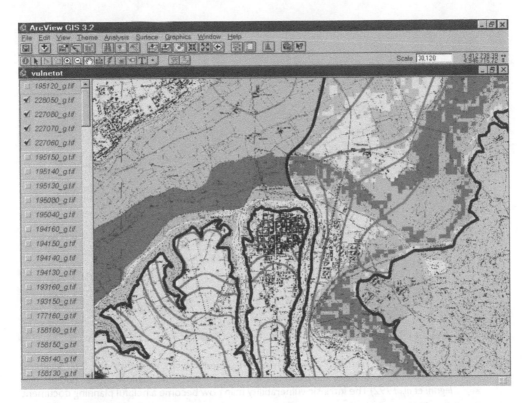

Figure 12.2 Vulnerability Map (extract) carried out using the combined approach. Vulnerability degrees: (Red) extremely elevate; (Orange) elevated; (Yellow) high; (Green) medium and (Cyan) extremely low. Isopiezometric contour lines (Violet). The black contour line mark the boundary between the two combined methods.

Contamination Sources) and many of the supply sources are concentrated (that is, *the subjects at risk – SAR*)? On the other hand, how can we carry out the evaluation of vulnerability and the risk to contamination for areas with great depth to water that can be described in less detail on the basis of *hydrogeologic situations and complexes?*

The solution that has been found for this problem and which has been tested, is the *combined approach*. This approach allows the GNDCI-CNR Basic method to be combined with the PCSM SINTACS method without continuity solutions: the latter in areas where the data that are necessary and sufficient to apply a parametric model exist; the first in areas where the great depth to water, the hydrolithologic and hydrostructural complexity and the lack of certain data on the terrains, the hydraulic conductibility and active recharge do not allow details to be obtained that can be compared with those that can be obtained using SINTACS (Figure 12.2).

4 Recommendations and conclusions

The combined approach allow to cover large and morphologically various landscape with an intrinsic vulnerability map. Moreover, to transform the document into an useful one for planning purpose, a number of discriminatory factors will be added (Figure 12.3).

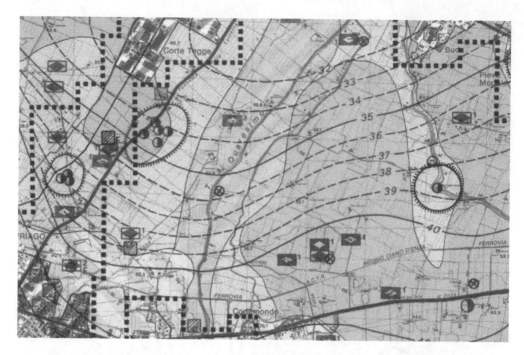

Figure 12.3 Overlying of CDP cadaster (several symbols), subjects to risk elements with their wellhead protection zone (in blue) and piezometric contour lines of the shallow aquifer (violet). *(Pellegrini et al., 1992)* The intrinsic vulnerability map now become a helpful planning document.

The first one is the contour field of the piezometric surface of the main aquifer which feed the tapping station of the surveyed area showing flow lines and other important feature, as groundwater divides, drainage axes, groundwater recharge zones, interfluves and so on.

The second is an overlay of all punctual and non-punctual contamination sources, between all the man-made pollution reducer and preventers (well-lined waste disposals, treatment plant of solid and liquid waste); the potential intake points of contamination material (active and abandoned quarry, mines, poorly-constructed wells, karstic depression and sinkholes (Civita and Rostagno, 2014) and so on); main subjects to pollution risk (groundwater extractions for drinking water supply, springs, well-fields and so on).

In this way, the map is ready for planning use, both at the prevision (and prevention) level and also to plan new potential dangerous settings, enforce land limitation of use, plan the structural and/or non-structural actions to mitigate the effect of groundwater pollution etc. Table 12.4 contains the main 10 cases of use for the vulnerability maps, integrated by the over presented information, for several degrees of the Authorities that work for water resources and environment conservation ad planning.

During 2004 the Vulnerability Map Italian Programme (VAZAR) and all the GNDCI-CNR was broken off cause the Civil Defense granting crisis. But after 2004 in a lot of Italian Regions, districts and municipalities, the vulnerability maps was drawn up for the same territories within the Water Protection Plan. The SINTACS method was adopted in very much cases although the BASIC Method was used in other cases. Only in some cases was used DRASTIC and other methods.

Table 12.4 Use of aquifer vulnerability maps for territorial and environment planning. C = Country; BA = River Basin Authority; D = District; CO = County; M = Municipality

Use of groundwater vulnerability maps	C	BA	D	CO	M
1. Forecasting of groundwater contamination by actual and/or potential point sources		•	•	•	•
2. Forecasting of groundwater contamination by actual and/or potential non-point sources		•	•	•	•
3. Previous evaluation of new territorial plans and transformations involving groundwater resources	•		•	•	•
4. Identification of suitable sites for settlements potentially hazardous for groundwater resources				•	
5. Evaluation of compatibility between existing activity and land use restrictions to enforce for prevention of groundwater contamination		•	•		•
6. Decision-making on new location of groundwater tapping works under contamination risk or indefensible by a protection zone system			•	•	•
7. Groundwater strategic resources, barely under contamination risk, to restrict previously (reserve areas)		•	•		
8. Aquifer decontamination priority identification (point and zone remediation plans)		•	•	•	
9. Groundwater monitoring planning to control, forecast and worn aquifer contamination		•	•	•	
10. Comprehensive environment planning of large areas to correct and minimize sever anthropic interference to groundwater		•	•	•	

Off Italy have drawn up the vulnerability map by Agency and State (Poland, Armenia, India, Slovenia, Mexico and so on).

References

Albinet, M. & Margat, J. (1970) *Cartographie de la vulnérabilité a la pollution des nappes d'eau souterraines.*Bull, Paris, BRGM s2, 3, 4. pp. 12–22.

Aller, L, Bennet, T., Lehr, J.H., Petty, R.J. & Hackett, G. (1987) DRASTIC: A standardized system for evaluating groundwater pollution potential using hydrogeologic settings. NWWA/EPA Ser., EPA600/287035. 455 p. 11 Maps and relate. Legends (with bibliography).

Andersen, L.J. & Gosk, E. (1987) Applicability of vulnerability maps. *Proceedings Int. Conf. Vulnerability of Soil and Groundwater to Pollutants, RIVM Proc. And Inf.* 38, pp. 321–332.

ANPA (2001) Linee guida per la redazione e l'uso delle Carte della Vulnerabilità degli acquiferi all'inquinamento. *Agenzia Nazionale per la Protezione dell'Ambiente – Manuali e Linee guida 4/2001*, Roma, p. 100, 1 CD.

Bachmat, Y. & Collin, M. (1987) Mapping to assess groundwater vulnerability to pollution. *Proceedings Int. Conf. Vulnerability of Soil and Groundwater to Pollutants, RIVM Proc. And Inf.* 38, pp. 297–307.

Born, S.M. & Stephensons, D.A. (1969) Hydrogeologic consideration in liquid waste disposal. *Journal of Soil and Water Conservation*, 24(2), pp. 52–55.

Breeuwsma, A., Wosten, J.H.M., Vleeshouwer, J.J., van Slobbe, A.M. & Bouma, J. (1986) Derivation of land qualities to assess environmental problems from soil surveys. *Soil Science Society of America Journal*, 50, pp. 186–190.

BRGM (1970) *Atlas des eaux Souterraines de la France*. éd. BRGM/DATAR.

Carter, A. D., Palmer, R. C. & Monkhouse, R. A. (1987) Mapping the vulnerability of groundwater to pollution from agricultural practice, particularly with respect to nitrate. In: Duijvenbooden, W. van, Waegeningh, H. G. van (eds). *Vulnerability of Soil and Groundwater to Pollutants. TNO Committee on Hydrological Research*, The Hague, The Netherlands. Proceedings and Information 38, pp. 333–342.

Civita, M. (1987) La previsione e la prevenzione del rischio d'inquinamento delle acque sotterranee a livello regionale mediante le Carte di Vulnerabilità. *Proc. Conv. "Inquinamento delle Acque Sotterranee: Previsione e Prevenzione", Mantova*, pp. 9–18.

Civita, M. (1990a) La valutazione della vulnerabilità degli acquiferi. *1° Conv. Naz. "Protezione e Gestione delle Acque Sotterranee: Metodologie, Tecnologie e Obiettivi", 3, Marano sul Panaro*, pp. 39–86.

Civita, M. (1990b) *Legenda unificata per le Carte della vulnerabilità dei corpi idrici sotterranei/ Unified legend for the aquifer pollution vulnerability Maps. Studi sulla Vulnerabilità degli Acquiferi*. 1 (Append.). Pitagora Editrice, Bologna. p. 13.

Civita, M. (1994) *Le Carte della vulnerabilità degli acquiferi all'inquinamento: Teoria & Pratica*. Pitagora Editrice, Bologna. p. 325.

Civita, M. (2000) Dalla vulnerabilità al rischio d'inquinamento. Relazione Generale. *Proc. 2° Naz. Conv. "Protezione e gestione delle Acque sotterranee per il III Millennio", Parma*, pp. 1–23.

Civita, M. (2005) *Idrogeologia applicata e ambientale (Applied and environmental Hydrogeology)*, CEA, Milano, p. 794.

Civita, M. & De Maio, M. (1997) Sintacs. Un sistema parametrico perla valutazione e la cartografia della vulnerabilita' degli acquiferi all'inquinamento. *Metodologia and Automatizzazione*, vol. 60. Pitagora Editrice, Bologna, p 191.

Civita, M. & De Maio, M. (2000) *Valutazione e cartografia automatica della vulnerabilità degli acquiferi all'inquinamento con il sistema parametrico. SINTACS R5 a new parametric system for the assessment and automatic mapping of groundwater vulnerability to contamination*. Pitagora Editrice, Bologna, p. 240 + 1 CD (with bibliography).

Civita, M. & De Maio, M. (eds.) (2002) *Atlante delle Carte di vulnerabilità delle regioni italiane*. DBMAP, Firenze, p. 366. Pubbl. n°2500 GNDCI-CNR.

Civita, M. & Rostagno, K. (2014) Le risorse dinamiche della Sorgente Sanità in Caposele (Sud Italia). (Dynamic resources of the Sanità spring at Caposele (South Italy). *Acque Sotterranee*, 3, 1/135, 9–24.

Civita, M. & Zavatti, A. (2006) *Un manuale per l'analisi di sito e la valutazione del rischio d'inquinamento*. Pitagora editrice, Bologna. p. 367.

Engelen, G. B. (1985) Vulnerability and restoration aspects of groundwater systems in unconsolidated terrains in the Netherlands, *Atti 18 Cong. IAH*, pp. 64–69.

Fenge, T. (1976) *Geomorphic aspects in aquifer vulnerability pollution, risk and protection strategy*. Western Geographical Service, Victoria, vol.12, pp. 241–286.

Foster, S. S. D. (1987) Fundamental concepts in aquifer vulnerability, pollution risk and protection strategy, *Atti International Conference Vulnerab. of Soil and Groundwater to Pollutants, RIVM Proc. and Inf.* 38, pp. 69–86.

Foster, S.S.D. & Hirata, R. (1988) Groundwater pollution risk assessment: A methodology using available data. *Pan-American Centre for Sanitary Engineering and Environmental Sciences (CEPIS), Lima*, p. 81.

Francani, V. & Civita, M. (eds.) (1988) *Proposta di normative per l'istituzione delle fasce di rispetto delle opera di captazione di acque sotterranee*. GEO-GRAPH, Segrate. p. 277.

Fried, J.J. (1987) Groundwater resources in the European Community, 2nd phase: Vulnerability – Quality (synthetical report unpublished).

Goossens, M., & Van Damme, M. (1987) Vulnerability mapping in Flanders, Belgium. Vulnerability of soil and groundwater to pollutant. In: Duijvenbooden, W. van, Waegeningh, G. H. (eds.). *TNO*

Committee on Hydrological Research, The Hague, The Netherlands. Proceedings and Information 38, pp. 355–360.

Haertle', A. (1983) Method of working and employment of EDP during the preparation of groundwater vulnerability maps. *International Association of Hydrological Sciences*, 142, pp. 1073–1085.

Hughes, G.M., Landon, R.A. & Farvolden, R.N. (1971) *Hydrogeology of Solid Waste Disposal Sites in North-Eastern Illinois, USEPA, SW122*. p. 154.

Josopait, V. & Schwerdtfeger, B. (1979) *Geowissenschaftliche Karte des Naturraumpotentials von Niedersachsen und Bremen, CC 3110 Bremerhaven Grundwasser, 1:200000*, Niedersachsischen Landesamt fur Bodenforshung, Hanover.

LeGrand, H.E. (1964) System for evaluating the contamination potential of some waste sites. *Journal of the American Water Works Association*, 56(8), 959–974.

LeGrand, H.E. (1965) Patterns of contaminated zones of water in the ground. *Water Resources Research*, 1(1), 83–95.

LeGrand, H.E. (1983) *A standardized system for evaluating waste-disposal sites*. National Water Well Association, Worthington, OH.

Marcolongo, B. & Pretto, L. (1987) *Vulnerabilita degli acquiferi nella pianura a nord di Vicenza*, Ed. Grafiche Erredieci, Padova. Pubbl. GNDCI-CNR n. 28.

OFRNARA (1994) Chapter I: EPA – Appendix A: The hazard ranking system. *Code of Federal Regulation, Protection of Environment, 40 CFR 300.A1*.

Olmer, M. & Rezac, B. (1974) Methodical principles of maps for protection of ground water in Bohemia and Moravia scale 1:200000, *Mem IAH 10*, 1, pp. 105–107.

Ostry, R. C., Leech, R.E.J., Cooper, A. J. & Rannie, E. H. (1987) Assessing the susceptibility of ground water supplies to non-point source agricultural contamination in South Ontario. *Atti International Conference Vulnerability of Soil and Groundwater to Pollutants, RIWM Atti and Inf.* 38, pp. 437–445.

Palmer, R. C. (1988) Groundwater vulnerability Map Severn Trent Water. *Soil Survey and Land Res. Cent.*, 8, p. 7. Carte.

Palmquist, R. & Sendlein, L.V.A. (1975) The configuration of contamination enclave from refuse disposal sites in floodplains. *Ground Water*, 13(2), 167–181.

Pavoni, J.L., Hagerty, D.J. & Lee, R.E. (1972) Environmental impact evaluation of hazardous waste disposal in land. JAWRA Journal of the American Water Resources Association, 8(6), 1091–1107.

Pellegrini, M., Tagliavini, S., Salsi, A., Voltolini, G., Zontini, S., Chesi, L. & Ganassi, G. (1992) *Carta della vulnerabilità degli acquiferi all'inquinamento. Alta pianura reggiana tra T. Crostolo e F. Secchia*. Studi sulla vulnerabilità degli acquiferi, 4. Pitagora Editrice. Bologna. p. 164.

Schmidt, R.R. (1987) Groundwater Contamination Susceptibility in Wisconsin. *Wisconsin Groundwater Management Plan Report No.5*.

Sotornikova, R., & J. Vrba (1987) Some remarks on the concept of vulnerability maps. *Atti International Conference Vulnerability of Soil and Groundwater to Pollutants, RIVM Proc. and Inf.* 38, pp. 471–475.

Subirana Asturias, J.M. & Casas Ponsati, A. (1984) Mapa de vulnerabilidad a la contaminacion de los acuiferos del valle bajo del Llobregat (Barcelona). *Metodo de trabajo y estudio de la evolucion de las extracciones de aridos*. Atti 1° Cong. Espan. de Geol. 1, pp. 795–809.

Trojan, M. D. & Perry, J. A. (1988) Assessing hydrogeologic risk over large geographic areas. Minnesota Agricultural Experiment Station, University of Minnesota, *St. Paul, Minnesota. Station Bulletin 585-1988* (Item No. AD-SB-3421).

Vierhuff, H., Wagner, W. & Aust, H. (1981) Die Grundwasservorkommen in der Bundesrepublik Deutschland. *Geol. Jb C30*, pp. 3–110.

Villumsen, A., Jacobsen O. & Sonderskov, C. (1983) Mapping the vulnerability of ground water reservoirs with regard to surface pollution. *Geological Survey of Denmark: Yearbook 1982*. Copenhagen, Denmark, pp.17–38.

Vrana, M. (1968) *Ochrana prostyc podzemnich vod v Cechach a na Morava. Vysvetlivsky k mape 1:500000*, Water Resources Plan Center, Praha (in Ceco).

Vrana, M. (1984) Methodology for construction of groundwater protection maps, Lecture for UNESCO/UNEP Project PLCE3/29, Moscow, September 1981. In: *Hydrogeological Principles of Groundwater Protection,* vol. 1, E. A. Kazlovsky (ed.), pp. 147–149.

Vrba, J. & Zaporozec, A. (eds.) (1995) Guidebook on mapping groundwater vulnerability. *IAH, International Contributions to Hydrogeology, 16 (1994)*, Heise, Hannover. p. 131.

Zampetti, M. (1983) Informazioni e dati relativi alla quantità e alla qualità delle acque sotterranee sotterranee nella Comunità Europea. *Inquinamento delle acque sotterranee da composti organoclorurati di origine industriali.* Monduzzi Editore, Bologna.

Zaporozec, A. (ed.) (1985) Groundwater protection principles and alternatives for Rock County, Wisconsin. *Wisconsin Geological and Natural History Survey, Madison, Special Report 8,* WI SR 8.

Chapter 13

The methodology of assessing groundwater vulnerability applied to Hydrogeological Map of Poland scale 1:50 000

P. Herbich, M. Woźnicka & M. Nidental

1 Introduction

Groundwater sensitivity is a complex issue, and this complexity is reflected in the large number of methods used in assessing groundwater vulnerability to contamination. Generally, two types of vulnerability are distinguishable: (1) natural vulnerability which is understood as a natural aquifer system property determining the risk of migration of contaminants from the ground surface to groundwater, and (2) specific vulnerability that also takes into account the type of the pollutant, its load, exposure time and the related spatial nature of the pollution outbreak (Żurek *et al.*, 2002; Witczak *et al.*, 2005; Herbich *et al.*, 2010; Duda *et al.*, 2011). Natural vulnerability depends on the natural hydrogeological conditions, including the infiltration of atmospheric precipitation, isolation from the ground surface, hydrodynamic conditions and percolation parameters of the aquifer and sediments in the vadose zone. An assessment of specific vulnerability can be made for individual pollution sources (Herbich *et al.*, 2007).

Since 2005, the Polish Geological Institute – National Research Institute (PGI-NRI) has been developing GIS database information layers for the Hydrogeological Map of Poland, scale 1: 50 000, to examination and characterise the first aquifer. In Poland, the implementation of the Water Framework Directive, which includes the determination of the potential for natural protection of groundwater, requires understanding of the vulnerability of the near-surface groundwater to contamination. It is one of the elements included in the risk assessment procedure on groundwater contamination (analysis of potential pressures and actual impacts on water status). In addition, information on the groundwater vulnerability to contamination is required for assessing the environmental impact of selected projects. Groundwater vulnerability is also taken into account when developing conservation action programmes and is used for the preparation of local development plans and in establishing intake protection zones and protected areas of major groundwater reservoirs. Accordingly, in 2006, the PGI-NRI launched a programme of development of successive GIS database information layers for the Hydrogeological Map of Poland for the first aquifer – characterising the natural vulnerability of the shallow aquifers to pollution. So far 390 map sheets had been completed (36% of the territory of Poland) by the end of 2014. It is planned to continue this programme to cover the whole country (Figure 13.1).

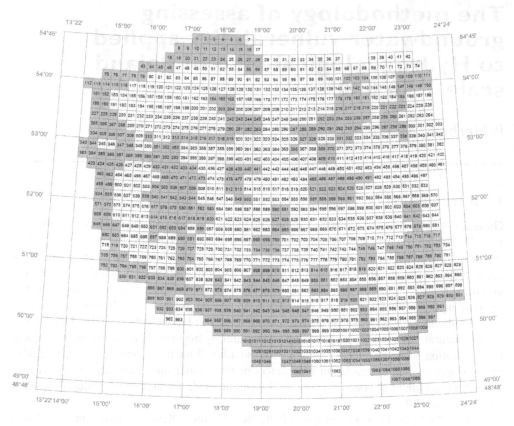

Figure 13.1 The current state of development of HMP information layers First Aquifer – groundwater vulnerability and water quality.

2   Methods for the assessment of groundwater vulnerability of the first aquifer to contamination

The GIS database information layers for the 1:50 000 Hydrogeology Map of Poland – First Aquifer Vulnerability to Pollution component, the methods applied for the preparation of The Map of Groundwater Vulnerability to Pollution, 1: 500 000 scale (Witczak, 2005; Duda *et al.*, 2011). These methods have been partially modified and adjusted for the more detailed scale of the new map. The method takes into account the nature of the terrain surface including land use, terrain gradients and soil types. Production of the 1:50 000 map sheets also required modification of the methods for industrial conurbations and areas with mine drainage issues.

Sensitivity classes of natural vulnerability, are based on mean residence time of infiltrating rainwater in the soils and rocks of the vadose zone for conservative pollutants migrating from the ground surface to the first aquifer (Herbich *et al.*, 2008). The mean residence time refers to the transport of all the water contained in the rock: in both active pores and pores of the rock matrix, as the value of the average annual infiltration rate (Witczak, 2005).

The algorithm of the assessment of the natural vulnerability of the first aquifer is as follows:

$$MRT = MRT_s + MRT_1 + MRT_2 \text{ [years]} \tag{13.1}$$

where:
MRT_s – Mean residence time in soil,
MRT_1 – Mean residence time in permeable sediments of the vadose zone,
MRT_2 – Mean residence time in poorly permeable and semi-permeable sediments of the vadose zone.

Mean residence time is determined separately for three layers that are characterised by different percolation parameters and are identified from a representative profile for the area (for explanation of symbols see Text and Tables 13.1–13.3):

1. For soil whose maximum thickness is determined as 1.5 m,

$$MRT_S = \frac{\left(1000 \cdot 1,5 \cdot w_{og}\right)}{R} \tag{13.2}$$

2. For permeable sediments in the vadose zone,

$$MRT_1 = \frac{1000 \cdot \left(\left(m_A - 1,5\right) \cdot \left(1 - S_p\right) \cdot w_{op}\right)}{R} \tag{13.3}$$

3. For poorly permeable and confining sediments in the vadose zone.

$$MRT_2 = \frac{1000 \cdot \left(\left(m_A - 1,5\right) \cdot S_p \cdot w_{oi}\right)}{R} \tag{13.4}$$

Both the type and thickness of the permeable and semi-permeable sediments of the vadose zone are averaged for the calculation areas, based on representative profiles. The coefficient of storage capacity of permeable and poorly permeable (w_{op}) and confining sediments (w_{oi}) are attributed to individual types of sediments in the vadose zone. Table 13.1 shows the values assumed for particular types of sediments. Percolation properties of soil are defined by the coefficient of storage capacity of soil profiles (w_{og}) determined from the Agricultural soil map (IUNG). The value ranges have been established in accordance with the Regulation of the Minister of the Environment of December 23, 2002. According to this Regulation, the protective capacity of soils results from the category to which the soils are classified based on their grain size composition (Table 13.2).

The proportion of poorly permeable and confining sediments (Sp) in the vadose zone profile is determined in percentages for 10-percent intervals, attributing the values from 0 (lack of poorly permeable and confining sediments) to 1 (complete isolation, confined groundwater). The assumed Sp coefficient value also involves perched aquifers (Sp value is

Table 13.1 Coefficient of storage capacity of sediments in the vadose zone: permeable sediments wop; poorly permeable and confining sediments woi* (volume percent) [-]

Rock medium type	Sediment type	Coefficient of storage capacity w_{op} and w_{oi}	
Permeable Fissured-karst	Limestones, dolomites	0,02	
Permeable Fissured	Granites, metamorphic rocks	0,01	w_{op}
Permeable Fissured-porous	Sandstones, flysch deposits	0,05	
Permeable Porous-fissured	Opokas, marls, chalk	0,05	
Permeable Porous, sand-gravelly	Variably grained sands, gravels	0,1	
Permeable Porous, fine-grained	Fine-grained, silty, argillaceous sands,	0,2	
Poorly permeable and confining layers – porous, clay-silty rock medium (k<10⁻⁶ m/s)	Loess, till	0,3	w_{oi}

Table 13.2 Coefficient of storage capacity of soil profile wog* (volume percent)[-]

Soil category	Soil type	Intervals	Average value within the interval
Very light	Podzol, lessive soil	0,11–0,14	0,12
Light	Muck and muck-like soil, brown earth, chernozem	0,14–0,21	0,17
Medium	Fen soil, rendzina, black earth	0,21–0,27	0,24
Heavy	Peats, silt-bog soils	0,27–0,46	0,36

* Regulation of the Ministry of the Environment of December 23, 2002 (Journal of Laws No. 241, item. 2093)

increased because pollutants reach the first aquifer with a delay). The thickness of the vadose zone (mA), which is one of the most critical factors affecting the result, is taken into account as the average value of the depth to the first aquifer (Table 13.3). This is one of the information layers developed for the Hydrogeological Map of Poland First Aquifer – occurrence and hydrodynamics.

The amount of infiltration (R) is defined from the classification of surface deposits in terms of lithological conditions and terrain gradient, degree of urbanisation and the so-called groundwater renewability index. The result depends to a large extent on the parameter defining the percolation properties of surface sediments (W – effective infiltration coefficient), which is determined from the Detailed Geological Map of Poland, 1:50 000, by assigning the values from 0.2 (very poor permeability of surface sediments) to 3.0 (very good permeability of surface sediments) to each lithological type.

$$R_1 = ZO \cdot W \qquad\qquad (13.5)$$

Table 13.3 Thickness intervals of the vadose zone

Depth intervals of the PPW occurrence (according to HMP FA-OH)		Value assumed for vulnerability assessment m_A [m]
<1		0,5
1–2		1,5
2–5		3,5
5–10		7,5
10–20		15,0
20–50		35,0
>50		50,0
For zww areas*	<5	2,5
	5–20	12,5

*zww – areas of diverse conditions of the first aquifer occurrence

where:
R_1 – recharging infiltration, excluding regions of poor percolation properties [mm/year],
W – effective infiltration coefficient [-],
ZO – groundwater renewability index FA [mm/year].

Construction of the 1:50 000 scale maps included areas of poor percolation and an effective infiltration reduction coefficient was introduced in areas where surface runoff is predominant:

In areas of dense urban-industrial development (U_1), including multi-storey housing, car parks and industrial housing – the infiltration value is reduced by a value between 70%-90%,

In areas of diverse urban development (U_2), including discontinuous urban development – the infiltration value is reduced by a value between 30% – 60%,

In areas where terrain gradients exceed 10° (U_3=1) – the infiltration value is reduced by half.

The infiltration values are reduced in areas where infiltration is substantially limited due to a high degree of urbanisation. The index U1 (infiltration value reduced by a value between 70%-90%) is applied for built-up areas with few green spaces (multi-storey housing, car parks, streets, pavements, industrial facilities, shopping centres), whereas the index U2 (infiltration value reduced by a value from the interval 30%-60%) is applied for areas of discontinuous urban development with a storm drainage system. Over the remaining area, the value of infiltration is not changed (U_1, U_2 and U_3 = 0). After correcting, the algorithm for determining the value of infiltration is given by:

$$R = ZO \cdot W \cdot (1-U_1) \cdot (1-U_2) \cdot (1-0,5 \cdot U_3) \qquad (13.6)$$

Based on the total mean resistance time of water in soils and rocks in the vadose zone, groundwater in the first aquifer is divided into five vulnerability classes (Table 13.4).

The information layer for the degree of groundwater vulnerability is a spatial layer with different colours for the individual vulnerability classes (Figure 13.2). Anthropogenic

Table 13.4 Vulnerability classes of the first aquifer

MRT [years]	Degree of vulnerability MHP-FA 1:50,000	Description*
<5	Very high	Vulnerable to most pollutants
5–25	High	Vulnerable to many pollutants, excluding strongly sorbed ones (e.g. heavy metals)
25–50	Moderate	Vulnerable to some pollutants, but only if they are introduced or leached off continuously
50–100	Low	Vulnerable only to conservative pollutants, introduced or leached off continuously, intensely and over a large area
>100	Very low	Not vulnerable to most pollutants

*after Witczak, 2005, modified

Figure 13.2 Information layers of the Hydrogeological Map of Poland, 1:50 000 First Aquifer – ground-water vulnerability – an example (map sheet 485)

objects and activities that may affect the groundwater chemical status of the first aquifer are presented against this background. In areas that are significantly transformed by human impact (industrial agglomerations, mine drainage zones) the dynamic changes the time of pollution transport to the first aquifer, and the resulting layers are modified by field observations. Vulnerability to contamination, understood in this way is determined from the current understanding of the first aquifer hydrodynamics.

The GIS database information layers of the Hydrogeological Map of Poland First Aquifer – Groundwater Vulnerability have been constructed in a sheet division using a topographical base at the scale of 1:50 000, coordinate system 1942. Digital exports are supported by the GeoMedia Access format and input to the integrated GIS database of the Hydrogeological Map of Poland 1:50 000 in coordinate system 1992.

3 Conclusion

The methods used to determine the degree of groundwater vulnerability to pollution enables its presentation in a uniform manner for the whole country. A multi-criterial analysis allows the variability of hydrogeological conditions of the first aquifer to be incorporated in the analysis. It also provides a basis for further, more advanced analysis of risk assessment, including consideration of various threat scenarios for groundwater. By 2014, 390 sheets of GIS database information layers had been completed. The work is ongoing.

References

Duda, R., Witczak, S. & Żurek, A. (2011) *Mapa wrażliwości wód podziemnych na zanieczyszczenie.* AGH, Cracow. ISBN978-83-88927-25-6.

Herbich, P., Nidental, M. & Woźnicka, M. (2007) Methodological guidelines of creating GIS database information layers of hydrogeological map of Poland 1:50,000 "first aquifer: Groundwater vulnerability and water quality". *XII Sympozjum Współczesne problemy hydrogeologii*, (2), 253–261.

Herbich, P., Ćwiertniewska, Z., Czebreszuk, J., Fert, M., Gej, K., Mordzonek, G., Nidental, M., Przytuła, E., Węglarz, D. & Woźnicka, M.(2008) *Wskazania metodyczne do opracowania warstw informacyjnych bazy danych GIS Mapy hydrogeologicznej Polski 1:50 000 "Pierwszy poziom wodonośny – wrażliwość na zanieczyszczenia i jakość wód".* PGI, Warsaw.

Herbich, P., Woźnicka, M. & Witczak, S. (2010) Hydrogeological cartography as a tool supporting water management, spatial planning and environmental protection. *Przegląd Geologiczny*, 58(9/1).

Witczak, S., Duda, R., Karlikowska, J. & Żurek, A. (2005) Możliwość wykorzystania mapy podatności do weryfikacji stref wrażliwych na zanieczyszczenia azotanami. *Współczesne problemy hydrogeologii*, 12.

Witczak, S. (ed.) (2005) *Mapa wrażliwości wód podziemnych na zanieczyszczenie 1:500000 (Plansza 1- Wody podziemne związane z wodami powierzchniowymi oraz ekosystemami lądowymi zależnymi od wód podziemnych; Plansza 2 – Podatność na zanieczyszczenie Głównych Zbiorników Wód Podziemnych (GZWP)).* Arcadis Ekokonrem Sp z o.o, Warsaw.

Żurek, A., Witczak, S. & Duda, R. (2002) Ocena podatności szczelinowych zbiorników wód podziemnych na zanieczyszczenie. In: Rubin, H., Rubin, K., Witkowski, A.J. *et al.* (eds.) *Jakość i podatność wód podziemnych na zanieczyszczenie.* Nauk o Ziemi Uniw. Śl, Sosnowiec, Wydz.

Chapter 14

Groundwater vulnerability assessment for the Hydrogeological Map of Poland 1:50 000 and Major Groundwater Reservoirs projects

K. Jóźwiak, J. Mikołajków, M. Nidental & M. Woźnicka

1 Introduction

There are interpretational difficulties related to the assessment of the possibility of protection of usable aquifers. In the Polish hydrogeological literature, there are many definitions related to the assessment of the possibility of migration of substances from the ground surface into groundwater. The terms: (natural and specific) groundwater vulnerability, resistance to pollution, degree of groundwater endangering and groundwater pollution risk – are often mutually mixed and sometimes erroneously interpreted. This is due to a variety of definitions of these terms. There is also a problem of the scale of a mapping project, as well as the knowledge on the area, on hydrogeochemical properties of substances considered as contaminants, and on the rocks which they percolate through.

The criteria adopted for the definitions often knowingly ignores the factor of time, reducing the calculations to static conditions specific to the aquifer system. Such simplifications are certainly possible in regional systems, but they should not be used in small open units, where the mean residence time is short. Additionally, these considerations need modification for large urban areas and small heavily human-impacted zones, e.g. large industrial plants.

A separate issue, in small-scale mapping projects, is the type of impact. In quasi-natural areas this problem can be neglected. However, in urban areas there are substantial problems associated with the transition from a spatial recharge (infiltration throughout the area) to a concentrated recharge, and with a significant reduction in infiltration. In these areas, there is an additional problem of local anthropogenic infiltration, which used to have been disregarded in the studies. For example, from 200 to 300 m3/h of water is introduced as a supplement to the water supply system of the left-bank districts of Warsaw (221 km²) during heating season. This amount, which comes from a leaky transmission network in the winter season, corresponds to a 4.0–5.9 mm column of water feeding the first aquifer.

Another problem is the definition of contamination/pollution itself. By analogy to the concept adopted for the needs of the Hydrogeological Map of Poland 1:50 000, and on the basis of the identification of potential or actual contamination sources, it can be assumed that a specific conservative pollutant can be applied to the environment, which will move into a usable aquifer after a set time. It can also be stated, based on the existing definition of groundwater contamination risk, that the risk level is a function between groundwater vulnerability, amount of load introduced into the environment, and standards assumed for the given type of environment.

In regional studies, introduction of specific types of contamination should be preceded by full hydrogeochemical assessment, i.e. an analysis of hydrogeochemical processes affecting water chemistry in the area should first be carried out. This allows a decision to be made whether the omission of hydrogeochemical analysis will enable reliable assessment results. It is possible to develop a simplified procedure involving the application of the concept of a water quality index – a value or specified parameter (concentration of the substance) determining the suitability of groundwater for specific purposes (usually for drinking), compared with the regulations or standards regarding water quality – indicative of anthropogenic transformation of natural composition of the water (Witczak and Żurek, 1994).

A groundwater contamination index is a change in physical parameters, chemical components and organic chemicals, suggesting water contamination. It includes specific indices (change in a single feature) and general indices (change in a certain group of features or parameters).

Regardless of the assumptions, the goal of every procedure (assessment of vulnerability, risk, resistance, etc.) is to apply it in order to prevent potential groundwater contamination. The first and the most important element is the water-bearing system, including:

[1] degree of aquifer confinement (variously defined);
[2] degree of aquifer recharge (direct rainfall recharge, lateral flows, inter-aquifer flows);
[3] hydrodynamic parameters of groundwater flow (and the resulting seepage velocity rate and groundwater flow rate);
[4] hydrogeochemical parameters (approved method of visualisation of the processes – advection model, sorption, half-life factors, ion exchange, etc.).

Another element is the identification of contamination sources and potential routes of recharge. In practice, each method uses different input assumptions. This affects interpretations carried out using different assumptions, even those based on the same data.

2 Methods: Hydrogeological Map of Poland and Major Groundwater Reservoirs

The Hydrogeological Map of Poland of Main Usable Aquifers (MUA), scale of 1: 50 000 (Paczyński, 1999), the protection of usable aquifers, uses five classes according to the degree of groundwater vulnerability. (Usable aquifer – an aquifer that should meet certain quantitative and qualitative criteria: thickness – >5 m, potential well discharge – >10 m^3/h, conductivity – >50 m^3/day, regional groundwater resource coefficient – >5 m^3/day per 1km^2). The rating of groundwater risk to pollution (from very low to very high) has been made with respect to the degree of aquifer confinement (a, ab, ba, bc, cb, c), Major Groundwater Reservoir protection zones, presence and quantity of contamination outbreaks (potential and actual), legislative protection of the area (groundwater intakes, national parks, nature reserves and landscape parks) and land use. In addition, the possibility of endogenic inflows needs to be taken into account, e.g. from sea water, high-mineralisation water (brines from the bedrock or a salt dome), as well as from aquifers with a high concentration of Fe, Mn and SO_4 (Paczyński, 1999).

Major Groundwater Reservoirs are determined in Poland as the most water-abundant structures, which are of strategic groundwater potential mainly for public water supply. The basic hydrogeological criteria for determining the Major Groundwater Reservoirs are: hydraulic conductivity greater than 240 m^2/day, potential well discharge greater than 70 m^3/h, possibility of developing a groundwater intake with a discharge rate of 10 000 m^3/day, and good water quality that allows for direct water supply or simple treatment using economically worthwhile technologies. In areas of significant water deficit, these criteria can be reduced, but they must highlight the water-bearing structures against areas with poorer parameters (Herbich *et al.*, 2009).

In accordance with the methods of determining protected areas of Major Groundwater Reservoirs (Herbich *et al.*, 2009), the protection of usable aquifers is implemented primarily through the designation of protected areas based on the flow time of water recharging the reservoir, indirectly understood as the time of migration of potential conservative contaminants from the ground surface into the aquifers that form the reservoir. This can be both the first (near-surface) usable aquifer recharged by infiltration of atmospheric precipitation and deeper-seated aquifers recharged either by percolation of water from the overlying aquifers or by lateral flow. The boundaries of the area that requires protection are determined from various criteria, e.g. from the map of vulnerability of protected aquifers. The construction of a map of protected areas is based primarily on the travel time of water recharging the reservoir (percolation through the vadose zone, inter-aquifer percolation, lateral flow, mean residence time) and the elements of land use and management. The rate of water migration into aquifers is determined from a mathematical model of groundwater flow.

Based on the assessment of travel time of water into the reservoir, the following areas are designated:

[1] highly vulnerable, with the travel time of water from the ground surface (the total time of seepage, percolation and lateral flow) delineated by an isochrone of <5 years;
[2] vulnerable, with the travel time of water between 5 and 25 years
[3], moderate and low vulnerable, with the travel time in the range 25–50 years; and
[4] very low vulnerable, with the travel time of water over 50 years. According to the adopted methodology, the reservoir's protection zone should include vulnerable and highly vulnerable areas.

One of the typical features of these methods is that hydrogeochemical elements are taken into account only to a small degree.

3 Assessment of the possibility of groundwater protection against contamination in the area of Major Groundwater Reservoir No. 148 Outwash Plain of the Pliszka River

In most of the area of Major Groundwater Reservoir No. 148, Outwash Plain of the Pliszka River, the main usable aquifer is represented by the Quaternary hydrogeological system. This system is composed of a near-surface aquifer and an unconfined (locally confined) upper inter-till aquifer. The main usable aquifer of the reservoir is unconfined; there is neither a lack of isolating poorly permeable sediments nor do they have other than a limited distribution and thickness (Kowalski *et al.*, 2011). The reservoir is recharged by direct infiltration. The lateral inflow from the outside is from the east.

The groundwater occurs in sand and gravel deposits of various origins, mainly fluvial and glaciofluvial. The usable aquifer is represented by variously grained sands and gravels, and gravelly-sand or glaciofluvial and fluvial sediments with a thickness from 10 to over 40 m. Their hydraulic parameters show considerable variations, the permeability coefficient ranges from 0.05 m/h to 10.0 m/h, and the storage coefficient varies (depending on the grain size) from 0.1 to 0.25. The hydraulic conductivity of the aquifer ranges from 0.8 m²/h to 113.0 m²/h, and is typically 5–20 m²/h. The potential well discharge rate is in the range of 10–60 m³/h.

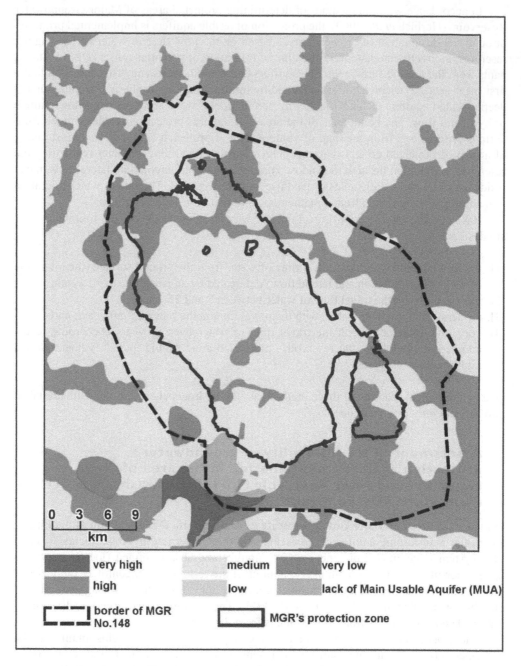

Figure 14.1 Major Groundwater Reservoir (MGR) No. 148 and the degree of groundwater vulnerability in the Hydrogeological Map of Poland 1:50,000.

The unconfined water table occurs at depths ranging from less than 1.0 m in the river floodplains and near the lakes, to more than 30 m b.g.l. in the watershed zones. According to the criteria adopted for the Hydrogeological Map of Poland, the aquifer confinement is poor and the thickness of poorly permeable sediments is up to 15 m.

Based on the methods adopted for the Map (Witczak, 2005; Herbich *et al.*, 2008; Duda *et al.*, 2011), four groundwater vulnerability classes have been identified within the reservoir (Figure 14.1). A substantial part of the reservoir is represented by highly and moderately vulnerable areas. These are mostly forested areas with a small number of objects potentially hazardous to groundwater, and with a poor isolation of the aquifer from surface contaminants. Areas of very high and very low hazard account for just a small portion. It should be noted that the degree of groundwater vulnerability within the protection area of the reservoir is high and moderate, as shown in the Hydrogeological Map of Poland.

According to the Major Groundwater Reservoir methodology, the starting point for determining the degree of groundwater vulnerability of the reservoir was the determination of percolation time through the vadose zone. This is one of the measurable criteria taken into account when assessing the natural groundwater vulnerability to contamination (Kowalski *et al.*, 2011). This time is calculated according to the Bindeman formula with the modification of Macioszczyk (1999). The calculation results show significant differences in the resulting time values, which derive from both the variable depth to the groundwater table and the presence of interbeds of confining strata in the vadose zone. The results vary from a few days to more than 25 years. Areas with the percolation time of less than 5 years are dominant. Only in the northern part of the Major Groundwater Reservoir does the travel time extends to more than 25 years, and this is associated with the occurrence of areas isolated by poorly permeable sediments and by thick overburden (Kowalski *et al.*, 2011). Figure 14.1 shows the extent of the proposed protection zone covering areas vulnerable and highly vulnerable, in which the travel time to the reservoir is less than 25 years.

The results of the assessment of to protect the Major Groundwater Reservoir No. 148 against contamination are significantly different depending on whether the Hydrogeological Map of Poland methods or Major Groundwater Reservoir methods are applied. Although the geological and hydrogeological conditions of this aquifer are simple, the results differ by an order of magnitude. These differences result mainly from the fact that the Hydrogeological Map of Poland methods involve the impact of potential contamination outbreaks on water quality when assessing the degree of vulnerability, which is neglected in the Major Groundwater Reservoir method. In addition, the identification of confining strata and poorly permeable sediments in the vadose zone is more thorough in the case of vulnerability assessment for the Major Groundwater Reservoir than for the Hydrogeological Map of Poland. It is difficult to apply both these approaches, as a reference point for the results.

4   Summary

The problem of assessing the possibility of groundwater protection against contamination, is demonstrated for Major Groundwater Reservoir No. 148 Outwash Plain of the Pliszka River. The methods used for the Hydrogeological Map of Poland and for the Major Groundwater Reservoiur show differences in the results, arising either from the criteria adopted for the assessment of the same geological and hydrogeological factors, or from the omission of anthropogenic factors. The differences cause significant interpretational difficulties, and thus can give rise to uncertainty as to the validity and accuracy of the assessment.

Another problem is the use of both products. In principle, both the Hydrogeological Map of Poland and Major Groundwater Reservoir are intended to be an element used by local administrative bodies. In addition, both of them are used in the development of Water Management Plans and Water and Environmental Programmes. Ultimately, at the stage of approval of the planning documentation, the resulting problems give rise to doubts at both national and EU level.

There is a need for developing uniform research methods that can be applied in the whole country area. Concurrently, the resulting outcomes must be clear and precise in order to be applied at an administrative level.

References

Database information layers of HMP 1:50 000, Warsaw. PGI-NIR.

Duda, R., Witczak, S. & Żurek, A., (2011) *Mapa wrażliwości wód podziemnych na zanieczyszczenie.* AGH, Cracow. ISBN978-83-88927-25-6.

Herbich, P., Ćwiertniewska, Z., Czebreszuk, J., Fert, M., Gej, K., Mordzonek, G., Nidental, M., Przytuła, E., Węglarz, D. & Woźnicka, M. (2008) *Wskazania metodyczne do opracowania warstw informacyjnych bazy danych GIS Mapy hydrogeologicznej Polski 1:50 000 "Pierwszy poziom wodonośny – wrażliwość na zanieczyszczenia i jakość wód".* PGI, Warsaw.

Herbich, P., Kapuściński, J., Nowicki, K., Prażak, J. & Skrzypczyk, L. (2009) *Metodyka wyznaczania obszarów ochronnych głównych zbiorników wód podziemnych dla potrzeb planowania i gospodarowania wodami w obszarach dorzeczy.* Ministry of the Environment, Warsaw.

Kowalski, J., Żaczkiewicz, M. & Szczepiński, J. (2011) *Dokumentacja hydrogeologiczna określająca warunki hydrogeologiczne w związku z ustanawianiem obszaru ochronnego Głównego Zbiornika Wód Podziemnych nr 148 Sandr rzeki Pliszka.* National Geological Archive PGI-NRI, Przedsiębiorstwo Geologiczne S.A. w Krakowie.

Macioszczyk, T. (1999) Czas przesączania pionowego wody jako wskaźnik stopnia ekranowania warstw wodonośnych. *Przegląd geologiczny,* 47, 731–736.

Paczyński, B. (ed.). (1999) *Instrukcja opracowania i komputerowej edycji Mapy hydrogeologicznej Polski w skali 1:50 000 (z aktualizacjami).* PGI, Warsaw.

Witczak, S. & Żurek, A. (1994) Wykorzystanie map glebowo-rolniczych w ocenie ochronnej roli gleb dla wód podziemnych. In: Kleczkowski, A.S. (ed.) *Metodyczne podstawy ochrony wód podziemnych.* AGH, Cracow pp. 155–180.

Witczak, S. (ed.) (2005) *Mapa wrażliwości wód podziemnych na zanieczyszczenie 1:500000 (Plansza 1- Wody podziemne związane z wodami powierzchniowymi oraz ekosystemami lądowymi zależnymi od wód podziemnych; Plansza 2 – Podatność na zanieczyszczenie Głównych Zbiorników Wód Podziemnych (GZWP)).* Arcadis Ekokonrem Sp z o.o, Warsaw.

Chapter 15

Irish groundwater vulnerability mapping and Groundwater Protection Schemes

Past, present and future

*M. Lee, C. Kelly, R. Meehan, C. Hickey &
N. Hunter Williams*

1   Introduction

Over the last number of decades, there has been increasing recognition and awareness of groundwater and groundwater protection, especially with the advent of the Water Framework Directive (WFD: 2000/60/EC; European Parliament and Council, 2000). Groundwater is a significant natural resource that currently supplies an estimated 20–25% of drinking water in Ireland (EPA, 2011), with the potential to supply more. It also provides significant contributions to wetlands and rivers, with an especially important role of maintaining flows through dry periods.

Groundwater in Ireland is protected under European Community and national legislation. The responsibility for enforcing the legislation resides with the Local Authorities and the Environmental Protection Agency (EPA). A vital element in groundwater protection is the use of relevant maps to make risk-based decisions. These maps and allied decision-making tools are provided in the Groundwater Protection Schemes (GWPS). A National GWPS, including a National Groundwater Vulnerability Map, was completed for Ireland in 2014. However, data gaps, the results of recent research, the future direction of WFD measures as driven by the EPA and the increasing need for higher resolution three-dimensional (3D) information all highlight that groundwater maps and conceptual understanding needs to be improved.

This paper charts the recent progress of the GSI's programme to provide Groundwater Protection Schemes and outlines the future areas of work that need to be addressed.

2   How a Groundwater Protection Scheme works

As shown in Figure 15.1, there are *two* main components of a GWPS:

- *Land surface zoning*: this is the general framework for a GWPS and is a map that divides the area into a number of *groundwater protection zones* according to the degree of protection required.
- *Groundwater protection response matrices for potentially polluting activities*: gives guidance on locating a specific, potentially polluting activity, depending on which groundwater protection zone the activity is (planned to be) in. The matrices describe: (i) the degree of acceptability of the activity; (ii) the conditions to be applied; and (iii) any investigations necessary prior to decision-making.

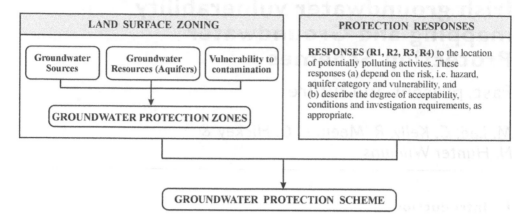

Figure 15.1 Summary of Components of a Groundwater Protection Scheme

3 PAST: development of Groundwater Protection Schemes

GWPSs in Ireland have been developing in their ideology ever since their inception in the mid-1980s. The underlying framework is based on a number of applicable, international methods used in countries with similar geology and hydrogeology. It also incorporates the results of Irish groundwater protection research (Fitzsimons *et al.*, 2003; Lee, 1999; O'Suilleabhain, 2000; Swartz, 1999; Wright, 2000). GWPSs focus on Irish water quality issues and potential sources of contamination, whilst being robust enough to adapt to emerging issues. The Groundwater Protection Schemes booklet, which outlined the methodology, was launched in 1999 (DELG/EPA/GSI, 1999).

During the mid-1990s, the GWPS land surface zoning maps were produced on a county basis as projects funded jointly by the GSI and the respective Local Authority for that county, who bore approximately 60% of the cost. For these Schemes, all three elements – groundwater vulnerability, aquifer, source protection areas – were mapped, assessed, and then combined on a county basis.

Throughout the 2000s, the work of the Groundwater Section in the GSI was strongly influenced by the needs of the Water Framework Directive. The GSI embarked on the provision of a National Aquifer Map in order to characterise Irish groundwater and delineate the "Groundwater Body" management units (WFD Groundwater Working Group, 2003). The National Aquifer Map was finalised in 2004.

During this period, the county GWPSs were still being produced and by 2007, 15 counties had been completed. However, the first of these schemes were already considered to be of less use because they did not use the most up-to-date data or methodologies, nor were they available in the required digital, GIS formats (Counties Clare, Limerick, Offaly and Tipperary South).

In 2007, the GSI received National Development Plan Funding, a portion of which was allocated to fund the National Groundwater Vulnerability Mapping Programme. The main driver was to provide the risk assessment layers necessary for the Water Framework Directive characterisation work, as well as provide one of the key layers for the GWPSs, which were required by all Irish Local Authorities. As such, the Local Authorities also funded a small proportion of the vulnerability mapping programme.

4   Groundwater vulnerability

The intrinsic vulnerability of groundwater depends on: (i) the time of travel of infiltrating water (and contaminants); (ii) the relative quantity of contaminants that can reach the groundwater; and (iii) the contaminant attenuation capacity of the geological materials through which the water and contaminants infiltrate. As all groundwater is hydrologically connected to the land surface, it is the effectiveness of this connection that determines the relative vulnerability to contamination. Groundwater that readily and quickly receives water (and contaminants) from the land surface is considered to be more vulnerable than groundwater that receives water (and contaminants) more slowly and in lower quantities (Daly and Warren, 1998).

In general, little attenuation of contaminants occurs in the bedrock in Ireland because flow is almost wholly *via* fissures. Consequently, the travel time, attenuation capacity and quantity of contaminants are a function of the following natural geological and hydrogeo-logical attributes of any area: (i) the subsoils that overlie the groundwater (most important); (ii) the type of recharge – whether point or diffuse; and (iii) the thickness of the unsaturated zone through which the contaminant moves.

The geological and hydrogeological characteristics can be examined and mapped, thereby providing a groundwater vulnerability assessment for any area or site. In summary, the entire land surface is divided into five vulnerability categories: Extreme (**X**) and (**E**), High, Moderate and Low, based on the geological and hydrogeological characteristics. The hydrogeological basis for these categories is summarised in Table 15.1.

Table 15.1 Vulnerability Mapping Criteria (Adapted by from DELG/EPA/GSI, 1999)

	Hydrogeological Requirements for Vulnerability Categories				
	Diffuse Recharge			*Point Recharge*	*Unsaturated Zone*
	Subsoil permeability and type				
Thickness of Overlying Subsoils	*high permeability* (sand/gravel)	*moderate permeability* (sandy subsoil)	*low permeability* (clayey subsoil, clay, peat)	(swallow holes, losing streams)	(sand & gravel aquifers <u>only</u>)
Generally <1.0 m	**Extreme (X)**	**Extreme (X)**	**Extreme (X)**	**Extreme (X)**	**Extreme (X)**
0–3 m	**Extreme (E)**	**Extreme (E)**	**Extreme (E)**	**Extreme (X)** (30 m radius)	**Extreme (E)**
3–5 m	**High**	**High**	**High**	N/A	**High**
5–10 m	**High**	**High**	**Moderate**	N/A	**High**
>10 m	**High**	**Moderate**	**Low**	N/A	**High**

Notes: (i) N/A = not applicable.
(ii) Release point of contaminants is assumed to be 1–2 m below ground surface.
(iii) Permeability classifications relate to the engineering behaviour as described by BS5930.
(iv) Outcrop and shallow subsoil (*i.e.* generally <1.0 m) areas are shown as a sub-category of extreme vulnerability and denoted as X. i)
(amended from DELG *et al.* (1999))

Generalised variations in subsoil thickness are interpreted using available depth-to-bedrock information, and knowledge of glacial erosion and deposition of subsoil sediments. Subsoil permeability ratings use a standardised, holistic approach that takes into account location- and subsoil-specific information such as textural classification and particle size analyses, as well as areal information such as geomorphic regions, recharge characteristic (drainage and vegetation), parent material information (*e.g.* bedrock) and topsoil information (Fitzsimons *et al.*, 2003).

As groundwater is considered to be present everywhere in Ireland, the vulnerability concept is applied to the entire land surface.

5   Mapping Groundwater vulnerability and providing GWPSs: 2007–2014.

By late 2007, a schedule for a five year mapping programme had been established and a consultant mapping team established[2] to undertaken work with the GSI, as well as a large number of GSI geological assistants and interns over the entire period.[3]

The programme was ambitious, with 15 counties requiring full vulnerability mapping within this time frame (see Figure 15.2).

In order to successfully complete the map production, as well as documenting the data analyses and work that supported the vulnerability classification decisions, a regimented annual routine was adopted. It comprised: 3–6 months of data collection and processing by the assistant geologists; 6–7 months of fieldwork by the mapping geologists (see Plate 15.1), which included a GSI drilling programme (to gather subsoil permeability and depth to bedrock data); 6–7 months of data interpretation and map compilation by the mapping geologists; Continual GIS work and support from the GIS consultant. There was also substantial technical input, supervision and support from the project manager, Quaternary specialist and GSI staff and considerable outreach and administrative support by the GSI.

Due to the success of the programme, the works were extended to re-map all or part of five additional counties (Shown in Figure 15.3). Although available in a digital format, it was now apparent that the data and methodologies used for these areas were out-of-date. To produce a *standardised* national product, additional work would be required. Unlike the previous four years' of original mapping, this work needed to be tailored to the specific needs of each county or area, which was both interesting and challenging.

6   Present: status and usage

Mapping, data analysis and map compilation work is now complete resulting in the National Groundwater Vulnerability (Figure 15.3), Subsoil Permeability, and GWPS maps, which is used by organisations and individuals. Their utility has been demonstrated by their inclusion in both national assessments and site-specific work *e.g.* planning and development, and research. All County Local Authorities involved now have GWPSs and are directed to use them in their County Development Plans by the Department of Environmental, Community and Local Government (Circular letter SP 5/03; DELG, 2003).

The National Groundwater Vulnerability maps (Figure 15.4) are used for source protection work throughout the country as well as being used as a national data layer *e.g.* EPA risk assessments for identifying problematic domestic wastewater treatment systems. They are

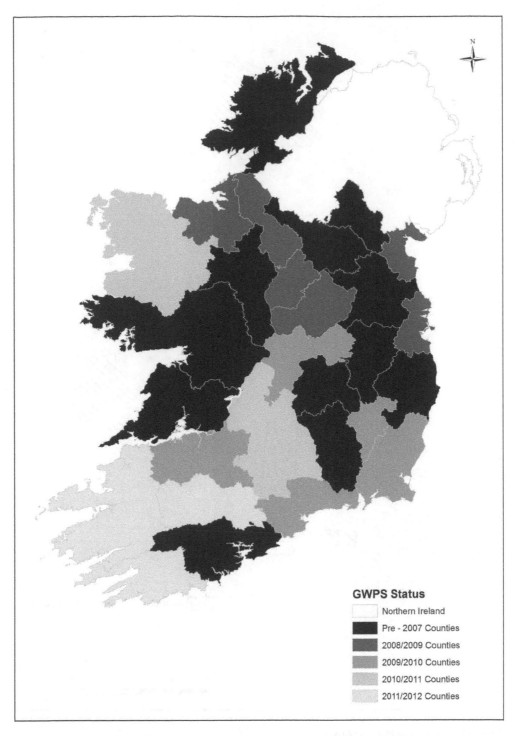

Figure 15.2 Areas in Each Mapping Year (pre 2007 until 2012).

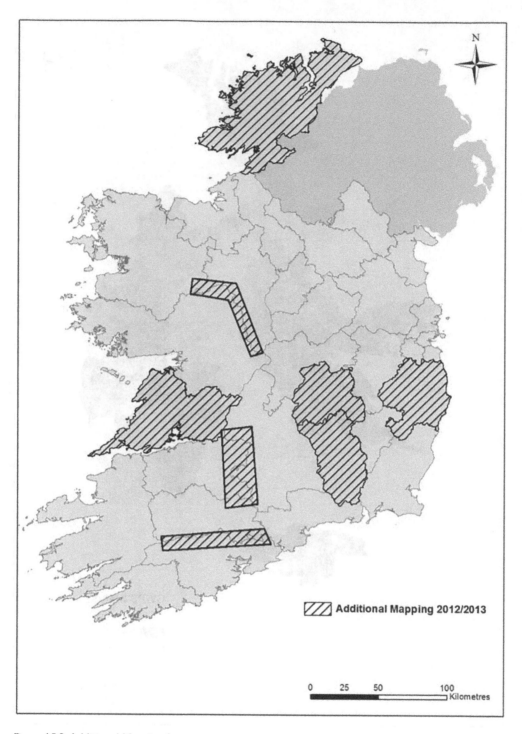

Additional Mapping 2012/2013

Figure 15.3 Additional Mapping Areas

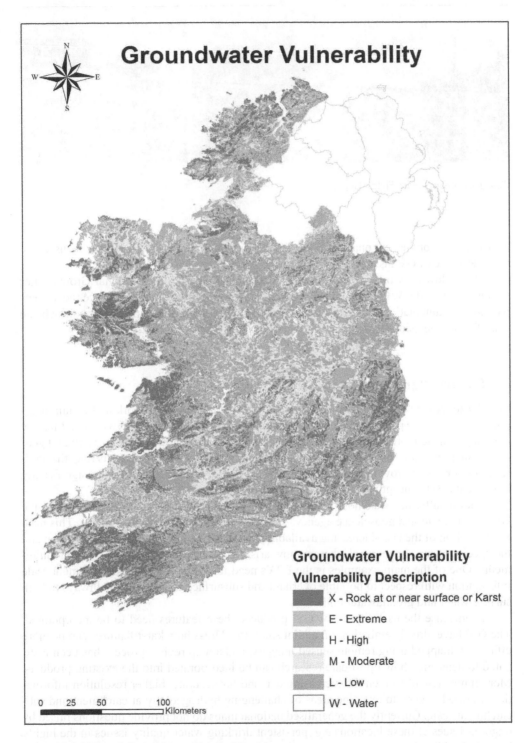

Figure 15.4 National Groundwater Vulnerability Map

Plate 15.1 Images from Field Mapping

also available for the next phase of WFD work, where proxy layers were previously used due to the absence of GWPS maps.

The follow-on work in 2014 has included amendments to the National Groundwater Vulnerability and GWPS Maps where newer data are available, and to produce the accompanying documentation and reports that outline the methodology and data on which the classification decisions were made.

fs

7   Future: "groundwater 3D"

In order to expedite the national GWPS map, the source protection work and certain mapping elements were postponed *e.g* labour-intensive and expensive countrywide karst feature mapping. Furthermore, during this period a number of relevant pieces of research and geological mapping have been completed, which could improve existing datasets (*e.g.* the aquifer map), or could provide additional data to layer on top of existing maps to significantly improve the 3D conceptualisation of the subsurface.

Additionally, our stakeholders now require higher resolution information, which the GSI – as the national geoscience agency – has an obligation to strive to produce. This need is a reflection of the ever-increasing availability of higher resolution digital information and the specific management needs that are now arising, especially through legislative requirements. One of the main examples is the EPA's need for more detailed, subcatchment scale information with respect to possible overland and subsurface contaminant pathways to both surface water and groundwater bodies.

To increase the usability of the GSI products, these features need to be incorporated. The GSI have already embarked on a pilot study to address how karst features can be more efficiently mapped using remote sensed imagery, and a map review process has been instigated to determine how up-to-date research can be incorporated into the existing products. More fundamentally, given the increasing demand for accurate, higher resolution information, the GSI is keen to examine areas of challenging hydrogeology at catchment and subcatchment scale. Currently the generalised national maps do not provide questions raised by ongoing issues at these locations *e.g.* persistent drinking water quality issues in the highly karstified midland region and understanding groundwater flow in the geologically complex area of North Cork.

The tasks ahead are daunting due to the size and quantity of questions that need to be answered. The GSI aims to work towards answers through the development of 3D models. In 2015, Geoscience Initiative (Government) funding was allocated to this process for an initial four year term. This has enabled a team of eight hydrogeologists and geoscientists to be put together to address these questions. The GSI is also utilising research funding programmes (*e.g.* the Irish Research Council's Enterprise Partnership Programme) and working with international partners (SISKA) who have advanced 3D modelling techniques in similar hydrogeological settings. The data collection, mapping and modelling programme for the next four years has a structure based on stakeholder needs, but is also flexible enough to develop in certain directions if required.

By the end of the four year term, the GSI will have improved local and national maps and conceptual models, and will be able to provide considerably more information on groundwater systems in priority areas. The programme will allow data to be collected, collated and analysed with the ultimate aim of creating 3D models that can add so much more to the understanding of our technical and non-technical stakeholders.

8   Conclusions

In Ireland, National Groundwater Protection Scheme and Groundwater Vulnerability Maps are now available. This is a huge achievement and it has taken over 20 years to get to this point. However, given the ever-increasing awareness of issues and data that are available, there are already improvements that can be made to provide better and more useful tools to meet environmental management needs. The GSI, through its new mapping programme and new and ongoing collaborations, aims to meet those demands.

Acknowledgements

Many individuals have contributed to the development of the GWPSs over the years. We specifically acknowledge the considerable contribution from Donal Daly (EPA, formerly GSI) for his development of groundwater protection concepts and GWPSs in Ireland. There has also been input from Geoff Wright, Vincent Fitzsimons, Jenny Deakin and Melissa Swartz (all formerly GSI), as well as Bruce Misstear, Paul Johnston and Dr. Eric Farrell (Trinity College Dublin). This paper is published with permission from the Director of the Geological Survey of Ireland.

Notes

1 Developed in conjunction with the Department of Environment and Local Government and EPA
2 Tobin Engineering Consultants: Coran Kelly (project manager); Monika Kabza, Orla Murphy and Melissa Spillane (mapping geologists); Talamhireland: Dr. Robert Meehan (Quaternary specialist and supervisor); Peter Cooney (GIS consultant).
3 Natalia Fernández de Vera, Jutta Hoppe, Magdalena Runge, Elena Berges, Shane Carey, Declan Kavanagh, Ramon Aznar, Rory Westrup, John Carroll, Marek Urbanski, Axel Keess, Sara Raymond, Nicola Salviani.

References

European Parliament and Council (2000) Water framework directive 2000/60/EC establishing a framework for community action in the field of water policy. *Official Journal of the European Communities*, L327, 1–73.

Daly, D. & Warren, W. (1998) Mapping groundwater vulnerability: The Irish perspective. In: Robins, N.S. (ed.) *Groundwater Pollution, Aquifer Recharge, and Vulnerability*. Geological Society, London, Spec. Publication 130. pp. 179–190.

DELG/EPA/GSI (1999) *Groundwater Protection Schemes*. Department of Environment, Local Government, Environmental Protection Agency and Geological Survey of Ireland, Dublin, Ireland.

Department of Environment and Local Government (DELG) (2003) *Circular Letter SP 5/03: Groundwater Protection and the Planning System*. DELG, Dublin, Ireland.

EPA (2011) *EPA Drinking Water Advice Note No. 7: Source Protection and Catchment Management to Protect Groundwater Supplies Version 1 Issued: 2 August 2011*. Dublin, Ireland.

Fitzsimons, V.P., Daly, D. & Deakin, J. (2003) *GSI Guidelines for Assessment and Mapping of Groundwater Vulnerability to Contamination*. Geological Survey of Ireland. p. 75, Dublin Ireland.

Lee, M. (1999) *Surface Indicators and Land Use as Secondary Indicators of Groundwater Recharge and Vulnerability*. Unpublished (Research) MSc Thesis, Department of Civil, Structural and Environmental Engineering, Trinity College Dublin, Dublin, Ireland.

O'Suilleabhain, C. (2000) *Assessing the Boundary between High and Moderately Permeable Subsoils*. Unpublished MSc Thesis, Department of Civil, Structural and Environmental Engineering, Trinity College Dublin, Dublin, Ireland.

Swartz, M. (1999) *Assessing the Permeability of Irish Subsoils*. Unpublished (Research) MSc Thesis, Department of Civil, Structural, and Environmental Engineering, Trinity College Dublin, Dublin, Ireland.

Water Framework Directive Groundwater Working Group (2003) *Approach to Delineating Groundwater Bodies: Guidance Note GW2*. Available from: www.wfdireland.ie.

Wright, G.R. (2000) QSC graphs: And aid to classification of data-poor aquifers in Ireland. In: Robins, N.S. & Misstear, B.D.R. (eds.) *Groundwater in the Celtic Regions: Studies in Hard Rock and Quaternary Hydrogeology*. Geological Society, London, Special Publications, 182. The Geological Society of London 2000.

Index

Bold page number refers to illustrations.

Series IAH-selected papers